JUST ADD WATER

The Realistic Guide to the Land, Landscaping, and Gardening in the Higher Elevations of the Great Southwest

Jim Koweek

Illustrated by: Zackery Zdinak

Additional Illustrations by: Dana Cude

Sonoran Wind Press
2007

JUST ADD WATER
The Realistic Guide to the Land,
Landscaping, and Gardening in the Higher Elevations of the
Great Southwest

© 2007 by Jim Koweek
United States ISBN # 978-0-6151-7465-5

Sonoran Wind Press
Whetstone, Arizona 85616

Summer Rain Photo by Meteorologist Christopher Reith
Cracked Earth and Arizona Poppy photos by the Author

All rights reserved. No part of this book may be reproduced in any form—except for the inclusion of brief quotes in review—without permission in writing from the author or publisher.

First Printing: November, 2007

In Memory of

Robert "Fitz" Fitzsimmons, as unique an individual as you could ever hope to cross paths with

And

Penny Artio, a good plant person and a gentle soul.

Dedication

This book is dedicated to Annette, Katie, Clay, Sommer, and Wesley.
And to the memory of my mom
Janice S. Koweek
who could have been a real writer
if she had chosen to.

Acknowledgements

I have always wondered why every book starts out with an "Acknowledgement" page. Now that I have put one together I understand why. There is no way this book could have happened without some real good help along the way. If I thanked everyone who has helped me the Acknowledgement Page would be longer than the text but here are some of the standouts. Most of these folks have endured my endless stream of questions for years.

The "Thank You" list starts with my first partner in a landscape business in 1978, Ronnie White and my old pal Paul Blackledge. We learned a little and laughed a lot. The University of Arizona Agricultural Extension Service is always a great resource. I especially want to thank County Agents Rob Call (Cochise County) and Dean Fish (Santa Cruz County) for all their help. Linda Kennedy, of the Appleton-Whittell Research Ranch, knows her grasslands and is an excellent teacher. Peter Gierlach and I always have some fun conversations. Sometimes they are even about plants. Sandy and Betsy Kunzer and Dale Armstrong have tried to teach me a little about geology. I am always saving questions for them. By now they probably wish I had taken an interest in astronomy instead of geology.

Thanks to Carol and Steve Schmidt, Joe, and the folks that work at High Noon Feed and Tack for being such good neighbors. "The Bulletin" and "The Farm and Livestock Trader" published my articles and put up with my seeming inability to recognize a deadline. By the way Peggy Dierking, while she was the editor of "The Bulletin", was the first one to ask me to write a column. Thanks for getting this whole thing started. Author Terry Mort encouraged me to put my stuff together in book form. Go buy all his books so he can afford to spend more time fishing.

I have learned lots about being professional and dealing with people from Rocky Harper. Don't ever stop picking Rock.

Finally, and most importantly, thanks to my wife Annette who serves as editor and grammar repairman. Without her there would be no book.

Introduction

Hi and welcome to Just Add Water: The Realistic Guide to the Land, Landscaping and Gardening in the Higher Elevations of the Great Southwest. This book, for the most part, is kind of a "greatest hits" collection of columns which appeared regularly in both "The Bulletin" (Sonoita, Arizona) and "The Farm and Livestock Trader" (Huachuca City, Arizona).

I hate to get bogged down with exact numbers but we are basically talking about the 3500'-6500' elevations. Those of us who live in this zone whether it is Arizona, New Mexico, Texas, or even Northern Mexico have a lot more in common with each other than we do with the low desert dwellers. This isn't a "How To" or "For Dummies" type of presentation, but it is meant to give a real feel for what makes working in this zone unique. Most of the information is based on what I have learned in over 25 years of working with plants and people in Southeast Arizona. Our goal is that you learn from some of these situations (my mistakes) and are able to save time, money, water, and frustration.

The driving force during this time has been the drought and its effects. You cannot overstate the importance of moisture in the Southwest. It always comes back to water.

Many of the specific examples that are used are local. This is mainly because I like where I live and don't care to get too far from home. You could personalize this book if you want by taking a marker and crossing out the towns and highways and then writing your own in their places if that makes you feel better. The information still applies.

One more thing before we get started, and you have probably figured it out already, so let's just get it out upfront. I make no claim to being a real "writer" I am more of a "talker on paper". Hopefully, either that is OK with you, or you have already paid for this book.

Jim Koweek November, 2007

Foreword

At first glance newcomers to the uplands of southeastern Arizona where I live may think they have arrived in the land of plentiful water. For goodness sakes, there are extensive grasslands and incredible oak woodlands. Maybe after climbing subtlety from the Sonoran Desert, and leaving behind the familiar saguaros and ironwood trees, the uplands look more like the moist hill country of Texas. But surprise, surprise, this is not the land of plentiful water. This is the land of 15 inches of annual rain if we're lucky and not in a drought. Welcome to the arid southwest where looks can be deceiving. This is the land of little rain where plants and animals have adapted to the lack of consistent water, and where for centuries human beings have tried to adapt. Sometimes with success, but more often that not, Mother Nature rules and things didn't work out.

If you are a gardener the uplands of the arid southwest become a test. It's still hotter than hell in the summer and the winters; well, temperatures easily drop to single digits and the cold season can extend far into April or May and start up again way too soon in October. And did I mention we don't get much rain?

All those plant and seed catalogs you get in the mail. You might as well throw them in the recycling bin the moment they arrive. This isn't New Jersey, this isn't Iowa and no, it isn't the hill country of Texas. You are going to need local help if you want to be a successful gardener in this arid land. You need someone that's "been there, done that". That's why this wonderful collection of essays will come in so handy.

Jim Koweek writes about landscaping, gardening and whatever come up in between, with a style that makes you think you're standing next to him He's looking you in the eye and he's telling you the way he sees it. He's honest, so if he blunders, he's going to tell you. And when he gets it right he's going to tell you. Best of all he tells with the wit and charm of a fellow that's lived in the southwest long enough to know what he's talking about.

Jim and I have been friends for almost 30 years. Go figure. But my point is that I write this with a little bias. Sadly for our wives we are both plants men, often finding ourselves discussing plants when we should be doing something that may make money.

So, now is the time to start turning pages and hanging out with my friend Jim as he takes you through the gardening seasons of the southwest uplands. You may want to crack a beer. You're going to be hanging out with a friend.

    Peter Gierlach (aka Petey Mesquitey)
    Near Pearce, Arizona spring 2007

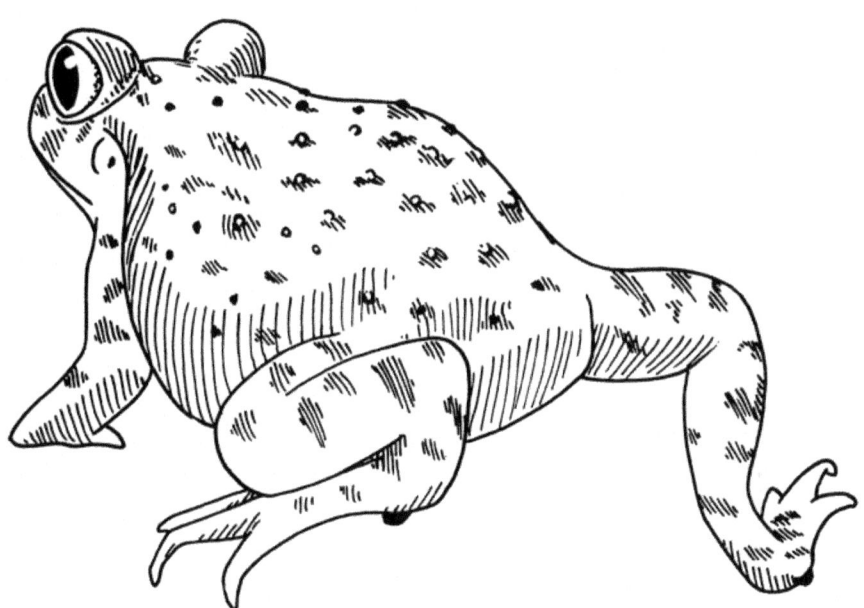

Prologue: The Five Season Year

Anyone who has lived in the upper arid lands of the Southwest for more than 365 days has probably figured out two things about our seasonal patterns. We may have historical averages for a particular season, like fall, but it is never a normal or usual fall. Secondly, and more importantly to us when trying to actually grow something, is that we enjoy a five season cycle instead of the more traditional four. Our seasons are based mainly on temperatures rather than a calendar. Here they are.

Summer is our most exciting season. Hot days and warm nights that are cooled off by the summer rains. Everything grows and comes alive. Gardens are productive and fruit trees start to bear. The native grasses green up for a while. If you are lucky, you can be stranded someplace for a couple of hours by water rushing through generally dry washes. It doesn't get much better than this.

Fall is wonderful. There is relief from the long summer's heat. Days are warm enough to get some work done but the nights cool off nicely. If there is any moisture it is probably driven by tropical storms in Mexico and can be good soaking rains. Hopefully you are picking apples and chiles.

Winter, like most places, is a time of dormancy. Temperatures, especially at night, vary based on elevation and location. Very few days fail to reach at least 32 degrees so we don't worry about a freeze line for our pipes. If it snows, enjoy it. Most of it will be gone by ten o'clock the next morning anyway.

Our next season is "Company." It is our first season of growth. Everything looks good. Fruit trees are blooming, penstemons are in flower, and all the native trees are putting on growth. This is good because it gives you something to look at and talk about with all your visitors from colder climates. The "Company" season usually falls somewhere on the calendar around late February and may continue into March and April.

"Too Hot for Company"*, our last season, usually kicks in sometime in late April or May. Temperatures can approach triple digits and humidity stays below 10 percent. May is statistically our driest month. Many plants, especially the natives, stop growing. This is our second season of dormancy. The "Too Hot for Company" season can be easily identified by watching for a big increase in the number of out of state RV's heading north on the interstate.

Now that we have correctly identified the five seasons, the rest of the book deals with how to work with them to get the most reward for the littlest effort.

Here's how to find them:

SUMMER	**3**
FALL	**39**
WINTER	**73**
COMPANY	**109**
TOO HOT FOR COMPANY	**139**

*I have worked with enough plant people to be aware of lumpers and splitters. For you splitters the "Too Hot for Company" season can be divided into "Too Hot and Windy" which is followed by "Just Too Hot."

SUMMER

WHERE ARE WE, AND IS IT EVER GOING TO RAIN?   p. 4

SO WHERE ARE WE?   p. 4

WATER YOU GOING TO DO ABOUT IT?   p. 6

IT'S GONNA DO WHAT IT'S GONNA DO   p.8

HOW GREEN IS IT?   p.10

THE SECRETS OF MANZANITA UNLOCKED   p. 12

PENSTEMONS ARE OVERRATED   p. 14

A TRUE STORY   p. 15

GOING WATERLESS   p. 17

IT'S ROUND-UP TIME   p. 18

THE ANT DANCE   p. 20

PARADES, PICNICS, AND HORNWORMS   p. 22

EXPERIMENTING WITH ALCOHOL   p. 24

SCIENTIFIC CONCLUSIONS ABOUT EXPERIMENTING WITH ALCOHOL   p. 25

HOT TIME ON MOUNT LEMMON, PART I   p. 27

HOT TIME ON MOUNT LEMMON, PART II   p. 29

SUMMER DUMBER THAN OTHERS   p. 31

TURN OUT THE LIGHTS, THE PARTY'S OVER   p. 32

WATTLE THEY THINKOF NEXT?   p. 34

SUMMERTIME, AND THE THOUGHTS ARE RANDOM   p. 36

Where are we, and is it Ever Going to Rain?

Geographically, geologically, and botanically most of us have more in common with the adjacent portion of Mexico than the rest of the US. We are the northern boundary for many "Mexican Species". Right now we are in of one of the worst droughts in modern history. Is this really a drought or are we just coming off of an 80 or 100 year wet spell? Danged if I know the answer, but after watching some summer clouds build one night I was moved to write this epic but sentimental poem:

> Red sky at night
> No stinking rain
> Red sky in morning
> No stinking rain either.

So Where Are We?

Two dirty looking hippies were coming off of a four day binge at the Arts and Crafts Fair. The older hippie looked up at the sky then turned to the younger one and said "Hey man, I am wasted. Is that the sun or the moon up there"? The younger one answered "I don't know man, I'm new in town".

Yes, I know that is an old joke but it is a good way to ask the question "So where are we"?

If you are reading this there is a good chance that you are, or want to be, in the Southwest portion of the good old USA. Even this little bit of information gives us a tip about our growing conditions. Expect hot summers and not much moisture. Most of the Southwest gets less than 20" of precipitation a year. This is the single biggest clue to figuring out what will grow here. Less that 20" means that not much plant growth is generated naturally so the soil doesn't have a lot of organic matter in it. When plants die here they tend to blow away. Our soil is various shades of alkali. Less than 20" also means not much humidity. The famous "dry heat". Leave the green and leafy, acid-loving plants like azaleas, rhododendrons, fuchsias, etc. to places where they will be happy - like Seattle.

There are four desert regions that make up most of the Southwest. They all have their own geography, weather conditions, and typical plants. I hate to break this to you but the background drawings, featuring saguaros (Sonoran Desert) and tall rock towers (Great Basin Desert) in the "Coyote and Roadrunner" cartoon are not anatomically correct. For our purposes we are mainly concerned with two of the desert types: the Sonoran and the Chihuahuan.

Despite common opinion, the Sonoran Desert is not the only desert in Arizona and New Mexico. The Sonoran Desert is typically hot in the summer and warm in the winter. Typical plants of the Sonoran Desert can take a lot of heat and drought but can't handle the cold. Think of saguaros and palo verdes as the poster children of the Sonoran Desert. Tucson and Phoenix are cities that are squarely in this desert.

The Chihuahuan Desert covers much of Southeastern Arizona and New Mexico. It has hot summers but cold winters (read, no bougainvilleas). Wind is a major factor here. The dominant plant form is shrubs. Most Chihuahuan plants, cactus and shrubs are built low to the ground and round so they can handle the wind. Willcox, AZ, Las Cruces, NM and El Paso, TX are all located in the Chihuahuan Desert.

The tough part is figuring out where these two deserts meet. Many maps show the Sonoran going to the west bank of the San Pedro River. It is the Chihuahuan on the east side. The truth is that these deserts don't meet, they overlap. These are natural boundaries not political ones. You can draw a map of specific localities for deserts and plants, but a as my old friend Dave Eppele used to say "too bad plants can't read".

Knowing your area is going to give you the best clue as to what type of vegetation will survive with minimum care. Moisture and cold are usually the most important factors. Here is a short list of average rainfall for some places in the Southwest. Remember that average technically means you never get exactly this amount.

Albuquerque, NM - 9.4"
Benson, AZ – 11.3"
Deming, NM – 10.7"
El Paso, TX – 9.54"

Globe, AZ – 15.9"
Las Cruces, NM – 11.4"
Oracle, AZ – 20.9"
Willcox, AZ– 12.2"

Water You Gonna Do About It?

I love contradictions. On one hand we have made recommendations to deep water your important trees to help them make it through this year's version of the continuing drought. On the other hand we have talked about the current drought and its impact on groundwater levels. You don't have to be one of those analytical geniuses to see just a little bit of a contradiction here. Do you spend the water to preserve the vegetation when water shortages might be just around the corner if the current conditions continue? This, my friends and neighbors, leads to a discussion of the most important, but probably least understood, subject in the Southwest: groundwater.

Ask four different people what our groundwater situation is and you will get at least four different answers, usually based on what the individual wants to believe. To a developer, groundwater is unlimited. To the person who favors "no growth", there is about six gallons left under the surface before we all dry up. To someone who is on a well for the first time, water is free; just use it, no cost. Quite honestly, most of us fall into the category of not even thinking about it until the well goes dry. I wanted to try and figure out what the real situation is so I turned to a hydrologist who has studied the water situation on a professional level, without political motives. In fairness to him, I shouldn't use his name because I draw my own conclusions from his set of facts. However, I would like to thank Dale Armstrong for an interesting and informative conversation the other day. Here are some things I have learned.

There is a tremendous amount of water stored in the valleys between the mountains. No one knows exactly how much. Most of it gets to the aquifer by passing through cracks close to the mountains. Melting snow is the main source of groundwater, not the run off from summer rains. In many areas drought is calculated by snow pack levels, and not rainfall. No one knows how long it takes the melted snow to make it from the mountains to your hose bib. Most of the so-called water problems are really well problems. Usage has lowered the water table to a level where the current well depth is no longer sufficient. You could think of our underground water and the material that holds it as a giant sponge where we have all chosen to sink our straws and drink. Suck too hard and you will affect the depth not only of your own straw, but of your neighbors around you too. There are heavy use areas in Sierra Vista where a state water inspector has told me that the water table has dropped over a hundred feet in ten years. This is important because the best quality water is closer to the surface. In some aquifers the deeper you go the more possibility of running into contaminants like lead or arsenic. This is one issue that we are all in together because groundwater doesn't recognize private or political boundaries.

Now getting back to the basic question: do we use the water in large quantities to do things like keeping vegetation alive? Or do we hoard it? The answer is yes to both. Use it wisely and make every drop count. This is a good time to start real water conservation. By cutting back on your water use in some areas you can still use it where it is needed, with no net increase in total consumption. It is like being on a water budget. Here is a very short list of water saving techniques.

Water Savers

- Knock a hole in your wall and run your washing machine drain outside. Each load of laundry can use up to forty gallons of water. That is enough to keep a few trees happy for awhile.
And, if you are building a new house, plan ahead to make use of this source for gray water. You can use the water from your showers, too. I am able to support a small home orchard with no additional irrigation.
- Shower with your old lady (or old man).
- Never wash a vehicle or a pet.
- Keep a bucket by the sink to collect small amounts of water that would ordinarily go down the drain. It might not be much, but it could keep a couple pots of flowers going.
- Mulch all your plantings. Research shows a good layer of mulch can cut water usage by almost a third.

Now it is your turn. We are in this one together.

It's Gonna do What it's Gonna Do

Here is a real news flash. We are still in a drought and so far the summer rains stink. I've heard all the predictions and so far most everyone has been wrong. The only people that have nailed it are the few that have said "it's gonna do what it's gonna do and there's not much we can do about it". Somebody understands. Here are a few predictions about the rainy season and explanations as to why they haven't panned out. You might notice the word "monsoon" is never used in this column. Somehow to me it just doesn't seem right. I'm hoping for summer rains.

"4th of July is the start of the rainy season". Oh well, I guess the rains didn't read the calendar this year.

"You need two hot weeks to bring the moisture up from Mexico" This theory is actually based on two different lines of thought. One is that two weeks are what it takes to start the convective current that transports moisture north from Mexico. When it is hot here the air rises and this lifting pulls up the moist air from down south. After a couple of weeks of movement a flow is created which brings in the rain.

The second group who backs the "Two Weeks" theory believes that rain isn't a result of a weather pattern. Rain is a direct result of suffering. If you haven't been miserable for at least two weeks you haven't paid the price. No free lunch. This group also believes that if there are early rains the season will be a bust. Again, not enough suffering. In academic circles this belief is called "Meteorological Calvinism".

"It takes a couple of really hot days to get things started" This is kind of the "Two Weeks Theory" condensed. We just had a couple of scorchers and no rain. As long as the wind is constant out of the north there isn't going to be any either. Perhaps we need to look into the possibility that the rains really aren't a result of a weather pattern.

Back in the mid 80's I worked with a fellow named Jose Samaniego who was from Douglas. While on a lunch break one day, Jose stopped chewing and started spitting on a Velvet Ant. When I asked what he was doing he told me that spitting on Velvet Ants made it rain. I pointed out that it might just be a coincidence that the ants showed

up in the rainy season. Jose said something to the effect that that was a dumb way to look at it and if I wanted it to rain I had better lugee up. Just to cover my bases, all my family, and I, has been spitting ever since. There is a scientific reason for including this story. This year I have not seen one Velvet Ant! No ants- no spitting. No spitting- no rain.

If you want to look out for these ants they are the ones with the bright orange puff balls on their abdomens. The rest of the body is black and they are about one inch long. We used to call them "Phyllis Diller" ants. Actually they aren't really ants but a wingless phase of a wasp. Remember, just spit on them and don't pick them up because they have a potent sting.

There you have it. Lots of hot air but no moisture. You can pray for it, do a rain dance, spit, shoot a cannon in the air, whatever you think might work. Just be aware, it's gonna do what it's gonna do.

How Green is It?

I decided that I wanted to get a better look at the countryside recently so I left the pick-up parked at the nursery and biked to work. It covers a distance of about 20 miles. Now for some this is a leisurely Saturday morning's ride. Not me. I am a driver or walker, not a cyclist. Real cyclists have a couple of things going for them. They have gear with gears, usually 15 or 20 of them to use. My ride is a combination of spare parts that I put together from parts of my kids' old Wal-Mart bikes. I used 2 gears on the trip. There were probably more but I wasn't sure if I trusted them. Also, when real cyclists get ready to ride, they climb in black shiny shorts lined with exotic animal fur on the inside. I prefer to have the animal parts on the inside of me. Blue jeans would have to do.

Good question to ask now might be "what the Sam Hill does taking a bike ride have to do anything? Isn't that a cheesy way to cob up a column?" Excellent question. The answer is that I would be observing the transition from the Madrean Evergreen Woodland to the Grasslands. I know this because I tagged along with my daughter Kate when she wrote about this subject for her class at the University of Arizona.

The hardest part is the first big hill that climbs out of the canyon bottom up to the highway. It is about half a mile long and steep. I believe it starts out at about an 89 degree slope. Gets steeper at the top. I made it to the top with only a mild feeling that I needed to puke. Now I was ready to observe.
It was green. Native grasses were at their peak. So were some of the invasives. There were at least 11 natives in bloom including 3-Awn, AZ Cottontop, Green Sprangletop, Vine Mesquite, Sideoats Grama, Blue Grama, Hairy Grama, Spruce Grama, Plains Lovegrass, Sacaton, and Wolftail. There were probably more but that is all I can identify by sight. Of course, representing the African invasives were Bermuda, Lehman's Lovegrass and Johnson Grass.

I have to admit that I really didn't notice much at 9 m.p.h. (average speed) that I hadn't seen at 35 m.p.h. (or so) in my truck. Smell and hearing are a totally different experience though. You can pick up the sweet smell of Desert willows in the canyon bottom at 5:30 am. The spicy fragrance of Cliffrose reaches out to you before you can even see them. Even squashed skunk smells better at less than 10 m.p.h.

Sounds are different too. The songbirds were really stretching out. Meadowlarks lined the fences by the Elgin School. There were lots of little brown birds that all look alike but sound completely different. This type of melodic experience is hard to notice when you have Hank Williams cranked up to 60 or 70 decibels on the truck radio.

Unfortunately, there were a few disappointments on the ride. I didn't find any good treasures along the way. Some of my best tools have been picked up along the roadside. I didn't see a one. There wasn't even a bungee cord to be found. When was the last time you drove someplace and didn't see one of those?

Another disappointment was the sunflowers close to Sonoita. I'll ask this question as politely as I can. Who is the moron that decided that they needed to be poisoned? The sunflowers in the late summer and fall are our reward for dealing with 5 months of wind, heat, and fires. They are a major food source for the doves. You are not doing us any favors by killing them. Don't do it next year. If I find out who makes these kinds of decisions I'll let you know and perhaps they can be contacted directly.

Overall I am glad I made the ride but I am definitely taking the truck home. To those of you that happened to stop by later that day and caught me napping on the seed sacks, I was dreaming of the next trip. No, it wasn't on a bike, it was on an ATV. And it had a big soft seat, lined with exotic fur.

The Secrets of Manzanita Unlocked

Most everyone takes special notice of manzanita. It triggers memories of hikes through hillsides for some. For others it reminds them of riding into thickets that were too dense to bushwhack. Whatever you think of manzanita, it's not doing too well now. It's being drought killed on a significant scale.

Our local manzanita (Arctostaphlyos pungens) is the point leaf type. There are three other species of manzanita in Arizona. All manzanitas are members of the heather family. That is probably why they don't handle this drought business too well. Generally they occur at about 5000' and above, on hillsides. They aren't too picky about soil.

Manzanita should be flowering now but they are not. Again, it is the lack of moisture. The flowers are white to pinkish and are often the only natural food for bees on warm late winter days. The most recognizable feature is its reddish bark. When dried, the wood makes great firewood.

About 10 years ago I made my living transplanting manzanita for nursery use. Must have done thousands of them. Usually we'd put about a 1' to 2' high plant in a 5 gallon nursery bucket. We tried all kinds of tricks to raise our success rate. Among other techniques, we pruned severely, watered before digging, and sprayed with an antidessicant to avoid dehydration. All of this seemed to have little positive effect. What did matter was the type of season we were having. In wet years, winter or summer, the success rate was probably about 75% to 85%. It went way down in dry years. I wouldn't even try digging this year.

Growing manzanita from seed is hard. I've only known two people who have had success. One was a nursery that is now out of business. They kept their secret to themselves and wouldn't tell anyone else how to do it. (One has to wonder if the "secret" was so valuable, how come they went out of business.) The other is my good friend Peter Gierlach. Peter claims that there is no secret. Just plant 1000 seeds and if you are lucky maybe 10 will germinate. Many people believe that the seeds have to go through fire to sprout. It's not that simple. Fire alone will not do it. Please don't go around setting fires just to try and get some manzanita started.

As a landscape plant, manzanita acts as a kind of living sculpture. The branching structure and color makes it a focal point. They work especially well against a light colored wall. Don't waste them as a hedge. There are better choices for that. Try Arizona Rosewood, Evergreen Sumac or some of the Texas Ranger types for that.

One thing I discovered about manzanita was that you can really push it with some extra water and fertilizer. Got over a foot of new growth a year. I also learned, the hard way, that this was a really bad idea. The new growth was especially tender and when fall came around every insect and fungal pest possible attacked it. This lesson should be applied to most native type plants in a landscape situation. Other than availability, there are two questions that are frequently asked about manzanita. The first is how do I acidify my soil so it will be OK to plant it in my yard? The answer is you don't. Most of the plants I dug were in soil of the 7.2-7.8 pH range. That is basic soil and it's probably what most of us have naturally. The other question deals with moving them out of their natural range. Manzanita with a little help (water) can make it most anywhere in SE Arizona. Years ago I sold some to be used in the median just north of the U of A on Speedway Blvd. If you go slow and look carefully you can see. Of course, going slow and looking carefully at plants in that area is not recommended.

Yup, manzanita is a neat plant in the right situation. Too bad they're going to be mighty scarce this year.

Penstemons are Overrated

We all like to see the spring penstemon blooms with their hot pinks, fiery reds, and purple spikes. Of course the hummingbirds really like them, too. But now it is June, the temperatures are in the mid 90's and with about 9% humidity, the question is: what have you done for us lately?

You sure were hot stuff in the spring after the winter moisture but guess what, so was everything else. It is easy to act like bigshots in the days of warm days and cold nights. What I really want to see are plants that thrive in the heat and are hitting their prime right now. Plants that are blooming and aren't fussy about soil. Penstemons are easy to over water and rot if the soil doesn't drain well. It would be nice if these plants could bloom through the fall during the hummingbird migration. I believe what I am looking for are plants called agastaches.

Agastaches (pronounced however you want because pronunciation of plant names is overrated also) are native plants to the higher areas of the Southwest that meet all the criteria that we mentioned above and more. Right now there is an agastache called Desert Sunrise which is in full bloom with 3' tall spikes of pink and orange flowers. Some people might even describe them as lavender and coral. Agastache rupestris is a native to Southeast Arizona and Southwest New Mexico. It is just starting to flower with orange flowers. This plant can grow anywhere from full sun to mostly shade. I have one planted in a heavy clay soil and it is doing fine. Best of all the leaves smell like root beer.

A few years ago someone gave me an agastache from the side of a creek by Rodeo, New Mexico. It had purple flowers. Not only did the hummingbirds, like it but after it set seed the yellow finches came to eat them. I guess they didn't eat all of the seeds because this plant came up on its own in several spots the next year. Ever seen finches on a penstemon?

I have to admit we are a little bit late to jump on the agastache bandwagon in SE Arizona. These plants have been a staple of New Mexico landscaping for a while. They got their big break a few years ago when somebody writing in one of those over slick magazines (might have been Sunset) wrote about visiting a garden in New Mexico where the hummers were ignoring all the rest of the plants and just feeding on agastaches. Considering the source, and that I am usually skeptical about claims like that, I basically ignored it. (Actually hummers can be very picky and will ignore plants that used to be favorites if there is another plant around that catches their fancy. They are definitely from the "what have you done for me lately" school of thought.)

In the interest of fairness there is a huge bloom of the native red penstemon (Penstemon barbatus) going on right now. We saw it all the way from Canelo to Glenwood NM. recently. That is a good plant which works in a variety of situations. It would work well planted with some agastaches. Yup, you could work in a few. Just remember, friends don't let friends plant too many penstemons.

A True Story

A friend stopped by the nursery the other day and was complaining that their tomatoes weren't doing very well. The plants had grown a bunch and lots of fruit had set but not many had ripened yet. Yes, the flavor on the ones they had eaten was good. How was my crop doing they wondered and did this seem like a good year for tomatoes? I had to admit I had been busy reseeding and doing lots of other projects lately so I hadn't even seen the garden for a couple of days. I would check and get back. When I got home I went to the garden and this is the truth.

The first plant I looked at was a Mortgage Lifter. This is an heirloom variety of tomato with a good local reputation as a heavy producer. Well it looked like someone had lifted all the leaves on the plant. Wasn't nothing left but stems with little bumps on them. Upon closer inspections the "bumps" turned out to be bugs. They were all less than one half inch long with a small heads, small midsections, and swollen bodies. They were probably swollen with my leaves. The body had a series of small black dots on the side. If disturbed, the bugs would roll up and play dead. Overall they were soft and squishy and kind of an ugly greenish-yellow booger color. Of course we picked them to get rid of them. I believe they were the larvae of the Colorado Potato Beetle. The adults have heads and midsections which are red with black dots. They are colored a lot like a Lady Bug. The larger body section is striped yellow and black.

Next up was an Early Girl tomato plant, usually one of the most productive varieties for this area. Taking hold of a branch which still had leaves on it, we searched for more Colorado Potato Bug larvae. We only found a few on this plant but when I looked down on the part of the plant I was holding, I saw that I was touching a small bright green Tomato Hornworm. We added this to the collection we were saving to give to the chickens. My research has shown that most chickens will eat small hornworms but leave the big ones alone. If you have lots of big ones use a Bt spray. We have covered this spray in other columns but it is a very safe and an effective way to get rid of these pests.

The next plant we looked at was an old-fashioned cherry tomato. The top of this plant was looking good with lots of new growth, flowers, and some fruit set. Unfortunately, the bottom wasn't looking too good. The very bottom had dead brown leaves, and right above that the foliage was turning yellow. The stems were a sickly drab green in that area. I am guessing that all this is being caused by Tomato Russet Mite. This pest is invisible to the naked eye but you can sure see the damage. The best way to treat this problem is with a wettable sulfur spray. This is a good low toxicity solution to the problem. I sprayed the other plants in the area as kind of a preventative measure. Besides, I had mixed up too much spray anyway.

So to go back to my friends question about "how does your tomato crop look this year?" I'd have to say at this point it looks normal, just like most other years.

Going Waterless

There are people who tell me that they enjoy the act of watering. It makes them relax and even has a meditative quality to it after their long day of stress. Personally, I think dragging a hose around is a big pain in the butt. That is one of the reasons that I like it so much when it rains. If it rains I get a break from watering. I am making this point to say that there is probably no one out there that wishes there was a formula for weaning plants off supplemental irrigation more than me. If you don't have access to water, plant cactus or succulents, put in some fake plants, or save your money and put in larger plants later.

Anything with leaves on it needs supplemental irrigation for a period of time? Yes, we are aware that the needles on cactus are modified leaves but we are talking "real" leaves here. Of course the question is how long is a "period of time"? I used to think I knew the answer to that question. A "period of time" was about 3 years. The first year you babied the plant, watering about once a week (more in the heat). The second year watering was reduced to once a week in the heat but every two to three weeks in the spring and fall. The third year the plant was irrigated enough to get through the little rough spot known as April 'til June. After that you and your plant would live happily ever after with no extra water.

That was the good old days. Then along came this little thing called the drought. (Just because we had some decent moisture this winter does not mean the drought cycle is broken.) You need to keep watering your plants as long as they need to stay alive. There is no formula to naturalize them. If you have limited water it is better to do a good job on limited plantings, than a rotten job on large scale plantings. If you can't keep your plants watered from April to June or July, don't waste time and money on them.

For those who have just bought a piece of property and are anxious to get started planting, here is a little insider's tip. Hauling water stinks! Wait until you have a well and somebody to apply the water before you plant.

Alright, I can hear the wheels turning. The way to get over on the natural order of things is to duplicate what is already growing. After all, those mesquites and oaks have survived on the natural precipitation. There is a big difference here. A 5' mesquite in the wild is probably within a couple of years of 8 years old. It has slowly been establishing a root system that is at least 15' in diameter to collect water. It also grew in a place where the conditions were favorable for it to survive. A nursery grown mesquite of 5' is about 3 years old. Its root system is 18" which isn't a whole lot to collect and store moisture for future use. It was put in a location favorable to the owner, not necessarily the tree.

The bottom line is there is no formula for going waterless. Don't plant it unless you can take care of it.

(A final note this week. While we are on the subject of long term success, congratulations to my good friends and very long term Sonoita residents Wayne and Clem Wright who recently moved to Tucson. This week they are celebrating their 70th wedding anniversary. Wayne, I am really thinking this one just might last.)

It's Round-up Time

Yup, we sure could use more rain. But since there has been some moisture (4"- 6" in most places) talk has changed. No longer are we complaining about the drought. Now we complain about the skeeters and the weeds. I can't do much about the skeeters but I have a hot tip for weed control.

Pull them. How's that for a hot tip? Hand pulling before they go to seed works for a lot of annual weeds. Tumbleweed and Pigweed

(Amaranth) fall into this category. Repeated mowing may also work on these kinds of weeds. This may take a period of several years to work.

Other weeds such as Russian Knapweed, Silver Nightshade, or Bermuda grass cannot be eradicated by hand. Because these types of plants spread from underground, hand pulling may actually cause them to spread faster. To control these type of weeds go chemical. It's Roundup time.

Roundup is a brand name for the chemical glyphosate. Other companies sell glyphosate but their brand name is different. They might even be cheaper. Killzall is one example.

A glyphosate works by shutting down roots. It is taken in through the leaf tissue, travels down the stem and kills the roots. The plant dies from the bottom up. With most plants this takes about 10 -14 days. With Bermuda grass expect about an 80% kill. You'll probably have to spray again.

Glyphosates work only if the plant is actively growing. Don't spray a dormant plant in the winter. If you really want to kill something and it hasn't been raining, water it a few times before you spray it. Roundup needs to sit on the leaves for about two hours without being washed off to be effective. If your leaves are dirty or dusty, hose them off before spraying.

Now here's the hot tip I promised, and thanks to Rob Call, Cochise County Ag. Agent, for this information. Research has shown that we're not getting all we should out of our Roundup applications. This is because the glyphosate molecules tend to lock up with the minerals in our water. To get the best results, mix some Ammonium Sulfate with your water BEFORE you add the Roundup. The minerals in the water will bind up the salts in the Roundup leaving the glyphosate free to do its job. I know it sounds strange to fertilize something that you are trying to kill. Use an equal amount of Ammonia Sulfate to the amount of concentrate that you are mixing in. These chemicals are not cheap so we want them to be as effective as possible.

A couple of other things about Roundup. It has no mammalian toxicity. In English, it doesn't affect people or pets at all. And for those of you that may be tempted to pour it on full strength, don't. You would just be wasting money. Roundup is as effective at the proper dilution as it is full strength.

The Ant Dance

I was listening to the radio the other day when one of my favorite commentators came on. No it wasn't Petey Mesquitey, it was Paul Harvey. Paul, better than anyone, is able to paint us a mental picture of the good times of summer. He was describing the first bite of sweet corn of the season. You could almost taste it.

I, too, am looking forward to the first bite of the year. It's not sweet corn though, it is red ants and they are doing the biting. Folks, this might seem a little weird but I kind of like being bitten by them. I don't believe I have ever been bitten any time of year but during the summer rains. When the ants bite it's the summer rainy season and the living is easy.

Ant bites can bring back some pretty good memories. These critters are really misunderstood so I'll try and give you some ant facts and back them up with "antecdotal" evidence.

First and most importantly, these are NOT Imported Red Fire Ants. They are Red Harvester Ants. The Fire Ants (Solenopsis invicta) are smaller, reddish brown, and their bite is much more painful. Fire Ants are mound builders. Last time I talked to the AZ. Dept. of Ag. we did not have any established colonies of Fire Ants in this part of the state. Our resident red ants are Harvester Ants (Pogonmyrmex sp.). They are larger than true Fire Ants, being about one half inch long. Their colonies are underground not in mounds. Nobody really knows how deep they go but it is estimated at up to 15' to 20' deep. The top of their colony is cleared of all vegetation so it looks like a bare spot 4'-10' in diameter. Harvester Ants are seed eaters. The young workers use their sharp powerful jaws to crack open the hard seeds. After their jaws become dull and can no longer crack the seeds, these older slightly worn out ants are demoted to "collector ants". So much for respecting your elders.

In the summer after a good rain, the ants swarm. You might have driven through these "flying ant" swarms. Young queens from one colony fly to nearby colonies to find a mate. They mate in the air, the male dies, and the young queen flies off to start a new colony. She may live happily ever after for up to 30 years.

Another misconception about these ants is their "bite". Technically, they don't bite, they sting. Not that it feels any different. The stinger can be left in the victim just like a bee. The venom acts on the lymph system and too much of it has the potential to cause problems.

I don't see these ants as much of a problem. It is the classic "don't bother them and they won't bother you". Sometimes it is hard not to bother them though. Several years ago I thought it might be amusing to dance the boogie-woogie on a colony. My kids were really amused when they watched me shuck my pants when I didn't get off the dance floor fast enough. The ants crawled up my legs and stung me.

The natural enemy of Harvester Ants is the Horny Toad. They will sit on the edge of a colony and gobble them up one after another. Last summer my trailer broke an axle right on top of a colony. I had to crawl underneath to work for a minute and get out and brush them off. Then go back under. What I would have given for a pocket full of Horny Toads. By the way these ants are edible. Just flick their heads off before you pop them in your mouth. Otherwise, as my son Wesley will testify, they might sting on the way down your throat.   Now maybe you will look forward to the first bite of the year.

Parades, Picnics, and Hornworms

Wherever you are, summer has its own rhythm. Wait for rain, go on a picnic, watch the parade, and look for hornworms. I never said all the rhythms were good. Hornworms are as much of our family traditions as playing summer baseball. I'll never forget the year we picked over 40 of them on just 14 plants in one day. I really don't like these critters. Maybe if I learned a little more about their lives I'd have a better appreciation for what I was squashing.

The first thing I learned is that what most of us call the Tomato Hornworm is actually the Tobacco Hornworm (Manduca sexta). The true Tomato Hornworm is more likely to be found east and north of here. The Latin name Manduca means glutton and the species sexta refers to the six spots on the abdomen of the moth that lays the eggs that become the Tobacco Hornworm. For some strange reason the common name of this moth is not the Six Spotted Glutton Moth but the Carolina Sphinx.

You may have already seen Carolina Sphinx Moths this year. They are easy to mistake for hummingbirds as they fly around looking for food. Sometimes they are called Hummingbird or Hawk Moths. At my house Four O'Clocks particularly attract them. The Carolina Sphinx lives as an adult for about a week then lays its eggs and dies.
The eggs are about 1mm in size and are a blue-green to yellow-green in color. If you see them on your plants pull them off and step on them. Eggs hatch in 1-3 days and the fun begins. It's hornworm time.

Hornworm caterpillars eat and eat. They are gluttons. Usually they are on our tomatoes but they will consume most any member of the Solanaceae family that includes peppers, potatoes, and eggplant. The typical hornworm is about 1"-3" long. It is tomato leaf green with a series of white stripes on its back. The most striking feature is a "horn", which looks like a curved spine, at one end. This horn is harmless and is used to scare off predators. The end without the horn is the head and the horny end is its butt. (Please no jokes here because this is a family column.)

These caterpillars are pretty hard to spot. It is easier to see where the damage (missing leaves) is and then search that area. If you see stems with no leaves and little black spots (caterpillar poop) you have hornworm problems. Hornworms don't bite or sting so if you see one pull it off and step on it. They must taste pretty rank because chickens usually won't eat them. Other controls include parasitic wasps and Bt. Bt is a very safe "organic" spray that kills caterpillars and nothing else.

After gorging themselves on your favorite plants the teenage hornworm begins to wander and look for a place in the ground to bed down. While looking for a place to sleep the hornworm often pukes up some of its last meals. It reminds me of some of those college parties you hear about. The teenage hornworm is about to become a pupa.

The pupa, which live underground, looks like a brown cigar butt with a string loop at one end. That string becomes the tongue. If you find a pupa, it is probably tempting to cut off the string. That would be cruel. Just step on the pupae and get it over with. Often just tilling up the soil will expose the pupas and end their life cycles. This stage can last from 3 weeks to 6 months. Then the Carolina Sphinx emerges and we get to start the whole cycle over again.

Maybe I shouldn't mention this but this entire mess could have been avoided. It seems that the Tobacco Hornworm was introduced in a load of tobacco from Nicaragua in 1641. If they had only taken care of business back then. Now this insect is gaining popularity. Science teachers raise them to demonstrate the life cycles of moths. You can buy eggs for your own experiments. In 1996, Tobacco Hornworms were sent on the space shuttle as part of the payload. Your tax dollars at work. They probably wanted to see how hard they were to step on in zero gravity.

Experimenting With Alcohol

(Jim's Note. I understand that there is responsibility when writing a column. There are times when you have to stop blowing smoke and put out some useful information that should be based on research and personal experience. I am not sure this is one of those times.)

This bit of research all started about two weeks ago when I went down to the tomato patch, the one planted in the old pig pens, to check for hornworms. The few hornworms that I found weren't a big problem. The blister bugs were. Blister bugs (aka blister beetles) are narrow, about ¼" wide by a little over an inch long. These were light brown but they also come in speckled grey and black. They move like they drank too much coffee. In ones or twos they are no problem, but a swarm of them can strip a garden in hours. Now take a guess what causes them to swarm. Ever been to a food court at a mall in the city? The same urges that drive teenagers to gather up there to eat bad food, listen to horrible music, and wear skimpy clothing are the same biological urges that cause blister bugs to swarm. Especially the part about the skimpy clothing. This was a major swarm and I didn't want these blister bugs acting like it was spring break in Puerto Penasco in my patch. I knew I had to do something or my crop would be set back several weeks in what is already looking like a slow year. There was one problem however. Some friends that I hadn't seen for a while were coming over to pick a few tunes. This meant that I had to make a decision, either deal properly with the blister bug infestation or play some music. So, as I was walking back to the house to tune up I noticed a couple of flycatchers perched on the fence. I hoped for the best. "Give 'er heck boys, it is the all you can eat blister bug buffet".

The next morning when I went down to check it didn't seem like the flycatchers had "given 'er heck" because the bugs were still there in the thousands. The tomato plants looked like someone had dragged a gravel rake through them. It was obvious that we needed to get some serious control going. But here was another fork in the decision making road. I didn't want to eat something that had been drenched in pesticides. On the other hand, if I didn't do something there wouldn't be anything to eat anyway. I was looking for a compromise solution so I decided to try insecticidal soap as a first resort. I could always get something stronger later if it was necessary.

Insecticidal soap usually works on soft bodied creatures like aphids. I wasn't sure if it would have any effect on the blister bugs. A homeowner can make their own by mixing a tablespoon of dish soap in a quart of water and spray away. The stuff that I had was a commercial brand and stronger than Ivory Liquid. The factory mixing instructions had fallen off so I decided to mix it at a rate of … Well, to be honest, let's just say I dumped a whole bunch in. I am not endorsing this dilution method but in this case we ended up with a pile of dead blister bugs. Mission accomplished.

With the blister bug invasion contained it was time to focus on public enemy number two, the hornworms. I still had some soap left over and was about to dispose of it in a safe manner, in compliance with state and federal guidelines of course. It wasn't like I was just going to dump it in the wash or anything. Then an idea hit me. The hardest part of hornworm control was finding them because they blend in so well. I didn't have to kill them; I just had to make them move. Then they would be easy to spot and I could use the ultimate biological control. Pick them off and squash them. Here was the chance for some bonified scientific research beneficial to hornworm sufferers all over the planet. I figured the best way to make them move was to irritate them. There still was some soap in the sprayer. After watching the blister bugs it was obvious that soap can be an irritating substance. Just ask any old hippies and about half the population of Europe. The basic question is would insecticidal soap make the hornworms dance the boogie? Another question is why is this article called "Experimenting with Alcohol"? I guess we will cover that next time.

Scientific Conclusions about Experimenting With Alcohol

We left off last time wondering if insecticidal soap would be irritating enough to hornworms to make them twitch and therefore be easier to spot. The answer is definitely not. Even when applied in a direct stream the hornworms seemed to look at you and say "thanks for washing off my food before I eat".

That could have, and probably should have been, the end of this scientific quest, but it wasn't. A few days later I saw some hornworms on a Desert Willow in my courtyard. Technically, I didn't see them at first. I felt some frass between my naked toes. (Because we are taking a scientific approach I am using the technically correct word, frass, instead of the common phrase, caterpillar crap.) Where there is frass, there are usually caterpillars. These weren't hard to spot. There were three hornworms, all about the size of a cheap White Owl Cigar, hanging upside down in the Desert Willow. They were big enough to be Longhorn worms. And they were just eating and pooping away. It was time to get going on the experiment, or again, getting technical, it was time to hose my toes off and crank up the sprayer. Alcohol was the next step. Alcohol kills many insects and really stings on open wounds. It can be really irritating. So can people that have consumed too much alcohol but that would be a different experiment. I put about 12 ounces of rubbing alcohol in a gallon of water. I chose to use 12 ounces because that was what was left in the bottle. We pumped up the sprayer and commenced firing.

Unfortunately it had no effect. Just to be sure I went down to the pepper patch. Nothing there either. So far it is hornworms 2 - Jim 0. Although as Tom Edison might say I had discovered two mixtures that didn't seem to bother hornworms.

I was about to retire the experiment again when it occurred to me that maybe I should try using something that was actually made for getting rid of them. In this case that meant using Bt, a bacterial spray that is designed for getting rid of caterpillars, and only caterpillars.

The spray rig was washed out and I carefully mixed exactly 1 tablespoon of Bt to a gallon of water. Sprayed them again. This time, they didn't seem to dance either. I went back and carefully dumped a whole bunch more in the solution. What followed was probably more like drowning than spraying. It was kind of like turning a 9 year old boy loose with a can of Raid. They still didn't move much but they looked dead when they fell out of the tree the next morning. Technically, I can't confirm that they were dead because the ants started to drag them off as soon as they hit the bricks. That is the way the experiment ended. The conclusion is that ants are a good biological control for hornworms. Sorry folks, but if I went on any farther, I might drive you to start your own experiment with alcohol.

Hot Time on Mt. Lemmon – Part one

Recently I was invited to spend a few days (and nights) on Mt. Lemmon as part of a reseeding project. Having never been involved in a project like this I decided to seek out the council of a couple of local experts as to what to expect for living conditions and hazards. Jim Maloney has dedicated his life to fighting fires. He probably has more experience "on the line" than most of the rest of us combined. His advice was kind of technical but certainly to the point. "Don't forget the Vienna Sausages". The next person who helped me out was our local jack-of–all trades and Tubac ambulance member, Brad Haber. Brad was pretty serious when he cautioned "Watch out for ash holes." I thought this was good everyday advice until Brad told me ash holes were pits of ash where trees had burned so hot that the stump burned underground. You don't notice them until you have stepped up to your knees in one. They are the second leading cause of injury to wildlands firefighters behind twisted ankles. Armed with this knowledge, my son Clay and I felt ready to climb the mountain.

Since we have returned, the question most often asked is "how does it look?" Of course there are two parts to the answer. For the most part the village and surrounding houses are toast. The plants and trees are better than you might expect.

What was houses is now pretty much piles of debris. Typically there is nothing left but the crumbling foundation, partially melted bits of metal, and ash. Almost every house has a few things that survived unharmed but there is no rhyme or reason as to what made it. A few cabins survived untouched. Again no apparent reason why they were spared. It is just the way it happened.

The village of Summerhaven and the surrounding area was being patrolled by troops (I'm assuming National Guard) for safety and security reasons. One of the guys on duty told me he had been in a lot of Third World countries and this is what they look like. Except, he added, in those countries people would be out picking through the rubble for anything of value. You can draw you own conclusions about that. My own is that even in situations of destruction we are still lucky to be living in the USA.

Judging by what melted on the hillside where we were camped, it was estimated that the fire must have been about 2000 degrees when it passed through. Exploding propane tanks may have had something to do with that. It was hot enough to ruin many of the foundations. If you kicked at the concrete blocks that make up the stem walls some of them would crumble. This is going to be a problem for those who want to rebuild but whose insurance doesn't cover foundations. Despite the losses most people we talked to seemed upbeat about rebuilding. Of course a few details will have to be worked out like asbestos removal, building materials, building codes, water issues, etc.

It seems that every extreme situation like this has some irony. If you walked down the hill toward Summerhaven there were only two buildings untouched. The Café was one. It opened for business on the Wednesday we were up there. I don't believe a cheeseburger ever tasted so good. The other was the Post Office. Now around here I'm told there used to be a Post Office in Canelo. Then it burned down. The Post Office was moved to Elgin where it burned down again. It seems that in the West there was some kind of reverse building code that said you had to burn down your Post Office every 30 or 40 years. So what survived the Mt. Lemmon fire? The Post Office.

Now out of the irony and into the fire. The mountain is in better shape than you might believe based on the media coverage, and it is already starting to come back. (To be continued next week.)

Hot Time on Mt. Lemmon   (continued)

Driving up Mt. Lemmon you wonder just how bad it is going to be. Not long into the journey you start seeing the first evidence of the fire. A burnt shrub or agave here and a black patch over there. Then, at about the 4500' elevation it turns bleak in a hurry. The area is kind of an oak – Manzanita – sotol - agave area. At least it used to be one. Now it is nuclear-looking scorched earth with the occasional blackened stump sticking up through the ash. It will come back. Might change some but it will come back. Remember the Ryan Fire last year? That fire burned the same sort of vegetation at a similar elevation and it is already healing nicely.

Keep climbing up the mountain and a funny thing starts to happen. After seeing the zone of destruction through the oaks I was expecting to see the top of the mountain look like something from the forest fire scene in Bambi. (I apologize for my use of intellectual film references to make this column seem more sophisticated.) Instead, the more we climbed, the better the mountain looked. First we noticed that some of the ridges or slopes were completely untouched. By the time you get to about the 7000'- 8000' level there is more green than black. At about 8000' you could drive for miles and might not even know that there was a fire. It's only when you reach the village of Summerhaven that the damage becomes apparent again.

I don't want you to get the idea that there are no dead trees along the way. There are lots of them but they died because of the drought, not the fire. The drought was what got us into this mess in the first place. If my memory is correct, 4 of the 5 biggest wildfires in Arizona recorded fire history have occurred in the last 18 months.

Back to the mountain. Some plants are already starting to come back. We saw lupines in bloom, and grasses sprouting. New Mexico Locust and other shrubs are growing back from the roots. Box Elders, a relative of the maple, are showing signs of green. Even the 2000 plus degree heat didn't kill all of the conifers. In the three days we were up there a few good rains fell. You could see the difference in the Southwest White Pines. Some of them are going to make it.

Helping out what is already happening is a large reseeding effort by both government and private sources. At about 5:30 AM we were overflown by crop-duster planes spreading seed. All day long helicopters carried large nets full of straw that they dumped over the seeded areas. The straw is supposed to hold moisture and help keep the seed in place. As it was falling, the straw looked like tan fireworks aimed back at earth. With the right kind of rain the mountain should green up this summer. No rain or a few violent microbursts and it is "oh well, we tried".

Will the mountain ever be just like it was? Heck no and I don't believe that you would want it that way. Part of the reason the fire was so fierce was 80 years of fire suppression. It was too thick with vegetation and houses and it paid the price. However, Mt. Lemmon will always be cooler, more forested, and a great place to escape the low desert heat for a day. Hiking and picnicking will return as soon as the mountain is opened to the public. People will probably be surprised at how much vegetation has come back already. In just 5 or 6 years people are already hiking on Chiricahua Mt. trails that they thought were ruined by a similar type fire.

Summerhaven itself will become a modern day boomtown. One contractor on the mountain had 3 months of work before the fire. He is now looking at 10 years of steady work. It will make turn of the century mining towns Duquesne or Charleston look like sleepy villages. It will sure be interesting, and fun, watching it come back.

Summer Dumber Than Others

We all know that some facts are kept secret. The 100 mile per gallon carburetor is kept off the shelf by Big Oil Companies and the discovery of intelligent life on Uranus has been covered up by the Government. They think we can't handle it. I happen to be aware of another startling, bonified truth which has been concealed but is about to be revealed. It is my duty to get it out in the public light. Can you handle it?

It is commonly accepted that when the internal body temperature rises several degrees above the norm of 98.6 fever starts affecting the mind and the thought process is warped. But did you know that when the external temperature exceeds 98.6 the same thing happens? The hotter it is outside, the dumber we act. Any prolonged temperatures above 102.5 degrees and we, as a regional population, turn dumber than a box of rocks.

I don't know why this fact has been kept hidden. Really we should be celebrating it. Give awards to the stupidest acts of the summer. Maybe even have categories. I was sure I could have won the "Dumbest Move in a Plant Related Field for Summer of 2006" for a couple of reasons.

Mostly this involved acts in my garden, which for various I didn't get around to planting until the middle of June. As we all know June is the time of year in the Southwest when I might actually want to go to the place where someone is always telling me to go just to cool off for the weekend. Now for stupid act number one. Because I just wanted to get it in, I didn't take any time to do any soil preparation. This left the soil very compacted from last year. And so when the rains came, instead of having good drainage, the soil held moisture at the surface and all my seedlings rotted. That is right; the heat had made me too dumb to grow zucchinis. But wait, it even gets worse.

I knew I had a lot of work-related projects going on and wouldn't have a lot of time for garden chores like weeding so I decided to go with a weed fabric. But instead of going with a real weed barrier that I had used before and knew worked well, I put down some coconut fiber erosion cloth. Yup, that is right. I used the exact same material for weed prevention last summer that I had used on several projects a couple of months earlier to hold soil and moisture and to promote the growth of the seed that had been placed underneath it. I blame that little decision on the heat. Maybe I couldn't grow squash but I had a blue ribbon crop of weeds under the cloth. Now you might have your own nominations but I am guessing that that example is hard to top in terms of heat related stupidity. (Maybe we should coin a new term of H.R.S. pronounced like Mrs. but with an H). However, just when I thought I had a lock on the Dumbest Move in a Plant Related Field, the phone rang.

On the line was George Stein, the owner of a small book publishing company in New Mexico who I had spoken with several times before. He had just gotten a call from a person whose wife was on the border of hysteria after she had gotten stuck with a cactus spine from a potted plant in their house. He wanted to know if it could be life-threatening. You could tell that the man had done some internet research because where else can you look to find information about life-threatening cactus attacks than book publishers? George wanted to know if I would be willing to help out and return the man's call. Obviously, this was way too entertaining for me to pass up. The first thing I did was to look up the area code to see where I was calling and it was Marin County in California. Big surprise there, huh? The number was answered by a professional sounding staffer that transferred me to the person who made the call. I explained to him that although many cactus spines may be irritating to the skin it was very unlikely that they will be lethal or even cause body parts to turn black and fall off. He thanked me for the information but then gave me another number and asked if I would call his wife and tell her the same thing. Why? Because if he told her what I said she would not believe him. It had to come from an outside source. . Now that really was the dumbest thing I heard all summer, maybe in any summer. Friends, I ain't no marriage counselor, but if your spouse won't believe you when it comes to the truth about life-threatening cactus attacks, you need to take some time off and get reacquainted as a couple. Go for a long ride in your car with the 100 mile per gallon carburetor and pick up your Award. I hope you find some intelligent life along the way.

Turn Out the Lights, the Party's Over

I was having a conversation with local astronomer (and fireman) Mike Shade the other day. Seems that both of us had reached the same conclusion, but for two totally different reasons. Porch lights are evil. They must go.

For connoisseurs of the night skies this is a pretty straight forward deal. Artificial light is a pollution that ruins the viewing experience. The more light there is at night the worse the viewing. There already has been a decrease in the quality of star gazing in the Southwest. Don't think it has happened? Go to town and try and spot the Milky Way.

I have noticed that porch lights make some other unwanted things happen. If you have lights, you will have bugs. One time my truck broke down late at night near the West Gate of Fort Huachuca. Near the guard shack they have a large watt spotlight that burns all night every night. The bug action there reminded me of one of those speeded up shots of downtown Tokyo. Bugs in turn will attract some other unwanted guests. They will attract predator bugs (read scorpions and centipedes) inside your house. And they aren't the only ones trying to eat the bugs you have attracted with your lights. Rodents like them, too. Now even if you think mice are cute the food chain doesn't end there because... rodents attract (you guessed it) snakes. So, cutting out the middle steps, porch lights attract snakes.

A little prevention can go a long way. Several years ago a gal that lived close to town told me that she had seen (and heard) five Mohave rattlers in her barn that summer. She wondered why she was being infested. I asked her how her feed was stored and she answered that she had a problem with rodents chewing through the paper and scattering it. In this case a garbage can for storage would have prevented the flood of Mohaves just like keeping night lights off can reduce the possibilities at your house.

Now I know what somebody out there is thinking "You'll get my porch light away from me when you pry it from my cold dead fingers. Them lights is what makes the house safe at night. If the bugs really are a problem, I'll just switch to a yeller bulb."

Nice try, but yellow lights don't eliminate or repel bugs; they just attract less of them. Besides there is another safety issue at work here in the Border States region. Folks that work for the Border Patrol have told me that porch lights act like a beacon for illegal border crossers. After spending days in the mountains, the light can really attract them if they want something like food, water, or use of a phone.

You might have noticed that I have never played the "energy card" as a reason to turn off your porch light. We all know it doesn't cost much to keep a 60 watt bulb burning for a few extra hours. At least monetarily it doesn't.

Wattle They Think of Next?

A few years ago some giant log looking objects started showing up on road side erosion control jobs. Our question for this week is what the heck are they and what good do they do? Hey, for a change these are easy questions. I'll start with the first.

The brown logs are called wattles. (I know most normal people think of wattles as the loose skin that hangs off of a turkey's neck, but use your imagination here.) They are either 9 inches or 20 inches in diameter and are made of straw, wrapped in netting. Most of the time wheat straw is used but in situations where there might be a concern about the weed potential of wheat, rice straw can be used

Wattles should be placed at a 90 degree angle to the flow of the water. Theoretically, they slow the water down and hold the dirt. Considering how much you paid for the dirt, I would think that holding it in place instead of letting it wash to the Sea of Cortez is a good thing. The pocket of soft dirt in back of the wattle is an ideal place for seed to germinate which is where the real erosion control takes place.

In order for a wattle to work it has to be installed correctly. That means digging them in to the ground about a third of their diameter and staking them. When they were first introduced, the crews would just lay them on top of the ground. Of course, if there was a good rain the water would undercut the wattle creating more erosion that if there was no wattle at all. Digging them in prevents this. Stakes should be pounded in every three to five feet.

At this point I know that you are probably gripping the edge of your seats wondering how you can get started working with wattles. The easiest way is to go and buy them. However I am guessing that there are some who will not want the "store boughts" but will want to roll their own. Just go get a bale of straw and some bird netting and twist some up. Don't try to use straw bales for this type of control. They don't let the water pass through so if there is a big rain event they will wash out.

Wattles aren't the only way to slow water down. If you are trimming shrubs or trees, lay the branches on the slope against the direction of the drainage. Staking them on a steeper slope is a good idea. Rock check dams can work wonders, too. The more of them installed, the better they work. Remember the lighter the material that you are working with, the less force it will take, so use the small stuff on gentle slopes.

A final word on wattles. You probably don't want to use them in areas that hold livestock. According to my old pal Jane Woods, she tried a few wattles in a pasture that was sheet flowing. A lead cow of her's took a dislike to the wattles and just lowered her horns and tore them up. Jane was thinking that the shape of the wattles might have triggered a defensive reaction. That might have been it, but my guess is that it has been so long since the cows have seen green grass that it just thought all that brown straw wrapped up was the world's biggest burrito.

Summertime and the Thoughts are Random

We are sitting on the cusp of summer and the intense heat that accompanies it. As the heat rises it elevates the cerebral processes of the brain. No wait a minute, I am not sitting on any cusp I am sitting on my butt in a rocking chair. And these aren't "cerebral processes" they are random thoughts because I am too lazy to research a real article or go outside and do any work. If there is a name after the thought, it's who I learned it from. The rest are more or less original. I hope a few hit home.

People who complain about the wind are the first ones to want a breeze when it gets hot and still.

Cicadas are the voice of summer heat.

"Look for rain 30 days after you hear the first cicada."
–Joe Quiroga

"It isn't really a "rain" if it doesn't run off of the roof"
-Wayne Wright

"It isn't really a "rain" if it doesn't penetrate the ground.
White people build on hills, Indians build down in the canyons."
- Phil Van Strander

Sometimes the most delicate, prettiest flowers have the most annoying seeds.

It is no problem to confuse oversimplification with common sense.

Declaring something "common sense" is easy once you know the answer.

"Scientific names should be used to make a discussion clearer, not confuse it." - Dave Eppele

Never buy apricots in a supermarket.

The best tasting tomatoes have a hint of "skunk" in their flavor.

By October you will know for sure what varieties of tomatoes did best this year. Unfortunately, that might not help you make good choices for next year.

A mesquite is no more drought tolerant than a cottonwood if it gets watered every day.

Pushing most native plants for maximum growth, with extra food and water, just makes them more susceptible to insect and fungal problems.

Want to attract wildlife? Plant something in an unprotected area.

"If you plant it, they will come." -Peter Gierlach

Knowing what caused a problem is a good idea before you decide to spray a plant.

Leafcutter Bees (the ones that cut circular holes in the leaves of roses, peaches, lilacs, etc) are good because they reduce the leaf surface during the hottest, driest time of year.

Land can't be returned to "the way it was" unless you have a specific date in mind. It is always changing. Sometimes the change is quicker than others.

Regardless of the drought, wildfires will always be a problem along the border until something is done about illegal immigration.

Some are sad when they see a dead tree. Others need the firewood.

There is no such thing as a "green thumb", you either water and pay attention or you don't.

Just because you are an old hippie doesn't mean you know squat about plants.

Never do today what can be put off.

In the West, it always comes down to water.

Folks, We are a lot better working with plants than we are with spelling and punctuation. If you see anything that need correction please contact us at jim@azreveg.com and we will try and get it straight. Who knows, maybe by the 847th edition we might have it all right.

                        Thanks, Jim

FALL

A SQUIRRELY PREDICTION   p. 40

IT AIN'T OVER   p. 41

MICROCLIMATES   p. 43

FALL TRADITION   p. 44

THE LAST WILD BLOOM OF FALL   p. 45

AN UNBELIEVABLE PLANT   p. 46

FRUIT TREES, PART I   p. 48

FRUIT TREES, THE SEQUEL   p. 49

APPLES—THE LAST WORD IN FRUIT TREES   p. 51

LET 'EM BEE   p. 52

LET 'EM BEE CLARIFICATION   p. 54

EXCUSE ME, YOUR GIRDLERS ARE SHOWING   p. 54

THE LAST WORD ON GRASSHOPPERS (WE WISH)   p. 56

MEANWHILE, BACK IN THE ORCHARD...   p. 58

ARIZONA CYPRESS BORER   p. 59

A SAD DAY IN SANTA CRUZ COUNTY   p. 61

DUMP REDEUX   p. 62

WE LOST A GOOD ONE   p. 63

WE JUST FLEW IN FROM MEXICO   p. 65

A FAIR PROJECT   p. 67

BACKPACKIN'   p. 69

PACKIN IT IN   p. 71

A Squirrelly Prediction

Any moron can look at a bunch of computer weather forecasts and come up with a long range prediction. It takes a very special kind of moron to come up with a long range forecast the way I did. It is all about the behavior of squirrels. In this case it is our local Rock Squirrel.

Rock Squirrels are pretty common around these parts. They grow to about 20 inches long with about half of that being tail. Their fur is a brownish gray color. Most of them live in holes in the ground. They eat a lot of juniper berries, cactus fruit, and nuts according to the books. I know I have seen them on the roads snacking on road kill, including other squirrels. I am guessing that behavior happens around mating season when you often see them perched on the highest branches of trees. They must not believe that bumper sticker that claims "vegetarians are better lovers". We might be getting off track here.

Back a little over a month ago we had some good October weather. There were warm days and cold nights. You could tell things were changing, getting ready for winter. Now here is the completely astounding part. In one week I had three different Rock Squirrel encounters. You probably had some around then too. First, one of those rodents moved in under the shack at the nursery. He started eating everything in sight. Plants, weeds, bags of seed, apples, and even pumpkins fell to his amazing appetite. During that same week I got a call from Patagonia resident Cici Finley. She wanted to know what would have made about a 4 inch diameter hole under a building where nothing had lived before. I didn't even have to see it to know that the Rock Squirrels were making a statement. Finally I was over at my friend John Everhart's house when he mentioned he might be having a squirrel problem. We looked around and sure enough there were squirrel teeth marks on the door frame.

Folks, I hope you have a lot of wood put up because all this squirrel activity means that we are in for a long cold winter. So far their predictions have been right on the money.

Last year we had a mild cold spell in October, and then it warmed up again for almost two months. It didn't really get cold again until right after Christmas when it suddenly changed. That was the most dramatic part of last winter as the rest was fairly mild. We even escaped a last killing frost which allowed us to have one of the best fruit crops in years. I don't remember even making a fire at our house until Thanksgiving. And you guessed it, we didn't notice any squirrel activity last year.

This year we are looking at a whole different story. For low lying areas the first hard freeze was in October. Early November brought some teen temperatures and this last front that passed through brought some of us down into the single digits. Winter has come hard and early. Our first fire was in October and we have been building fires morning and night for several weeks now. The squirrels predicted this, and according to them, it will continue.

By the way, according to official weather sources a "hard freeze" is considered several hours of 28 degrees or less. I don't believe these weather experts know Jack Frost about hard freezing. When you wake up and can't get water for coffee because your pipes are frozen, that is a "hard freeze". Even squirrels know that.

It Ain't Over 'Til It's Over

Mid November in Arizona. Sure is beautiful, isn't it? NO IT ISN'T! It's too dang dry. I am not talking grumpy, never satisfied with the weather, rancher dry. It is a measurable, compare it to other great droughts of the past, type dry. Too quantify this drought we use something called the Palmer's Drought Index.

In the 1960's, Wayne Palmer developed a scale to measure long term drought conditions. He started by averaging temperature and precipitation for a local area. Then he compared current conditions for that area over a period of time, usually months, with the average. The results were graded from −4 to +4. A rating of −4 is "extremely dry" while +4 is "extremely moist". Currently we are rated between a −2 and a −3 which is "moderate" to "severe" drought. We are getting worse by the week. The key to the Palmer Index is that it is based on extended periods of time. One good rain or snow helps but doesn't break a drought.

If you want to learn more about the Palmer's Drought Index get a copy of Gary Gruenhagen's article in the Cochise County Master Gardener's Newsletter. He really explains the methodology clearly.

Here's what our −2 (moderate) to −3 (severe) rating means to us and what we need to get back to "average". Please cut me a little slack. I am working with recorded averages. We all know someone whose neighbor got 157" of rain last summer but claims they didn't get any. Most of us got about 6"-9" of moisture. That is 30% to 50% below usual. We would need to get 8"-12" in the next six months just to get back to a "O" on the Index. That is 150%-175% of what we average. I know the percentages don't line up exactly with the amounts, but remember that temperatures are a part of the Palmer Index also. Immediate relief would come in the form of about a 2" rain. That number is getting bigger by the week.

In recent history, 3 of the 4 last winters have been near record dry. Very unofficially, we've gotten about 2" of moisture in 7 or 8 months during that time. Our oaks and junipers are amazing in their ability to cope with dry times but are beginning to show the stress. There are spots on western and southern slopes in the foothills of the Huachucas that have lost significant numbers of trees. If this arid trend continues, some of our oak woodland will probably become grasslands and the grasslands will look more like desert scrub (picture Highway 90 between Huachuca City and I-10). I've already seen stands of Chihuahuan Pine at lower elevations really take a beating from the stress of drought. Change happens quicker than you think.

Ready for some good news yet? Droughts come and droughts go just like the extremely wet years do. The really bad droughts like the ones in the 50's and the Dust Bowl 30's lasted about 5-7 years. We've measured climate scientifically for about 100 years. Before that, you use tree rings, pollen, and other methods to get a feel for the climatic history of the last 2000 years. Overall weather patterns haven't changed much in that period of time. Some times are better than others. Current predictions for this winter are for wetter than normal, thanks to "El Nino". The change could kick in sometime in December. If it does happen, it won't take long for people to complain about the cold and wet. They will probably long for a return to the warm, sunny beautiful days of November.

Microclimates

The first cold spell of the year finally arrived. Took a while to get here, but the temperatures were almost as cold as I've recorded in several years. At the nursery in Sonoita it was 12 degrees and at home it reached 3. I'm guessing that that is colder than most places in the Southwest with similar elevations. Before you go suggesting that I get some new thermometers, let's look at why.

Both the nursery and my house are in low places compared to the surrounding area. Cold air is heavy and it sinks. Washes and low spots collect the cooler air and hold it. The larger the collecting area the lower the temperatures. At the base of the mountains, one morning I measured an 8 degree difference in less than a 12 foot rise in elevation on a south facing slope.

We seem to pay more attention to elevation and not enough to the situation. A 5000' hilltop is going to be warmer than a 5000' wash in the same basic area. South and west facing slopes receive more sun and are warmer. The wind generally comes from the southwest so the east side is protected. Each of these conditions creates a microclimate. Microclimates are just small areas that act differently than the surrounding area. You can use them to your advantage in your home landscape. Many cactus and succulents thrive on those hot western exposures. Green and leafy plants usually benefit from a windbreak. Put them on the east side. Late frost ruins an entire fruit crop. If your ground permits it keep them out of the low spots.

Microclimates can be artificially created. A shade tree creates a cooler protected area underneath. A large boulder absorbs heat in the day and radiates it at night. This could be the ticket to survival of some marginal cactus such as the Golden Barrel. And, the area directly around a water feature is more humid.

Even as I write this I'm creating an "internal microclimate". I stoke up the fire and there is a zone of warmth by the old wood burning stove. I better have another cup of coffee, move closer, and think about it some more.

Fall Tradition

No, I haven't been drinking. The reason that my truck swerves going down the road is that I'm trying to squash grasshoppers. It's one of my favorite fall traditions. It's a feeling of accomplishment. But who knows, by the time my youngest are grown the roads might be too crowded to practice safe swerving. Maybe flattening grasshoppers with a motor vehicle might be against the law. Either way another piece of Southwestern Americana would be gone.

On a brighter note, another positive thing you can do with your time in the fall is to plant. According to the University of Arizona, fall is our best time to plant. Their reasons are pretty straightforward. Putting the plant in the ground now when things are cooling off eliminates stress. You probably won't see much top growth, but until the ground cools off, the plant may be producing roots. This means your plant will be ready in the spring to take advantage of 100% of the growing season next year. With some care, your fall plantings will be established enough to deal with the heat and wind next spring and summer.

There are a few exceptions to the "fall is best" theory. Plants that freeze all the way to the ground sometimes don't get enough start to make it through the winter. Mexican Bush Sage and Coral Bean are a couple of examples. Cactus and succulent cuttings should also be planted in warmer and drier times. They have a tendency to rot in the damp. Also be wary of plants that are not acclimated to our area. Actively growing stock from low desert nurseries can be damaged in one of our early cold snaps. Of course we have the reverse situation in the spring.

Probably the biggest obstacle to fall planting is that it just somehow feels funny. We seem to be internally wired to plant and begin growing in the spring. We'll do this even though we know our late frosts are legendary. Fall planting really does make sense.

Here's a tip that might make fall planting a lot more practical, especially if you have lots to plant. Hire someone to dig the holes. You can still do the actual installation. The dirt goes back in a lot easier than it comes out. It's money well spent. You might even take a fall drive while the work is being done.

The Last Wild Bloom of Fall

It's fall and most everything in the wild is shutting down or turning brown. The one exception is Rubber Rabbitbush, which is now in its glory. It grows in colonies along washes and is covered in small yellow blossoms.

Rubber Rabbitbush is known by a handful of names. It's leaves and stems do contain a small percent of rubber. In New Mexico, where it

is used extensively as a landscape plant, it is called Chamisa. Locally, like any gray-green plant that grows in low areas, it is sometimes called sage. Botanically speaking it is Chrysothamnus nauseosus. The plant grows 4-5 feet tall and grows mainly along washes.
Last Wednesday I decided to get a closer look at the Rubber Rabbitbush.

From about 40 feet away I could tell the place was really humming. There were hundreds of bees, both native and honey, just covering the flowers. Upon closer inspection my untrained eye was able to pick out 8 different species of butterflies. There were probably more. It's a good pollination strategy to be the only plant blooming when the insects are still active. Kind of like being the only guy at a dance.

Rubber Rabbitbush is easily confused with another gray-green plant that grows near wash bottoms and has yellow flowers, Thread-leaf Groundsel (Senecio Douglasii var. longilobus).
It can be important to tell them apart because the Groundsel can be toxic to both horses and cattle. It shuts down their livers. Thread-leaf Groundsel has a fleshier leaf and flowers in the spring. It is a shorter plant, only getting about three feet tall. Rubber Rabbitbush can be browsed with no problems.

Rubber Rabbitbush could be used in a home landscape to extend the fall blooming season. Mix it with some of the other plants that might still be blooming now, like the salvias. Give it good drainage or it may rot.

An Unbelievable Plant

There is a plant out there that just gets more positive response in the nursery than almost all others put together. By the way it excites people you might think it has memory enhancing, weight loss potential, aphrodisiacal, cholesterol reducing, muscle building, and hair restoring properties. Of course the only plant on earth that could generate all this buzz is rhubarb. That's right; the legendary high cuisine of the frozen north seems to have cult status here in the Southwest.

Bring up rhubarb and I can guarantee you from personal experience that the first response that you will get is stories from someone about all the rhubarb they ate in the summers of their youth. I usually don't have the heart to point out that where rhubarb grows abundantly they were probably so happy not to have 3 feet of snow on the ground that they would have eaten tumbleweed. Next come the stories of the strawberry-rhubarb pies. I've made those. They weren't too bad except for the rhubarb part. I tried making an apple-rhubarb pie and basically ruined the apples. Yup, I have been told that you need "lots of sugar" when you cook with rhubarb. Why you can even eat it plain if you dip it in "lots of sugar". Folks, you could probably eat dried, white dog poop if you dipped it in "lots of sugar". I think I'll pass.

Taste aside, the question is can you grow it here in SE Arizona? In the right location it does real well. Make sure it is planted in well drained soil. A break from late afternoon sun is a good idea, also. My experience is that rhubarb is much more drought tolerant than you might expect. Don't use gray water to water it though. Years ago I watered a couple of plants with gray water and by the afternoon they were shriveled and it looked like someone had let the air out of them. They took about two months to come back to their previous state. When they were looking good again I got to wondering if it really was the gray water that affected them. My scientific curiosity got the better of me and I tried it. It really was the gray water.

In some ways rhubarb is a good crop for these parts. Since you eat the stalks, late frosts aren't a problem. For you rhubarb elitists - get used to green stems. I have tried several different varieties and none of them holds the red color once the season warms up. I have gotten one apple crop in about 11 years but every year we have rhubarb out the ears. Lay off the leaves. They may look like edible greens but they contain oxalic acid. In the period after World War I people tried to eat them as a source of greens and many of they got sick. Fortunately the concentration of the oxalic acid is low enough that you would have to ingest about 11 pounds for a lethal dose. If the leaves taste anything like the stalks I don't believe that is going to be an issue, even with "lots of sugar".

All right, to be fair rhubarb can't be all bad. Helene Wingert told me that back in Holland rhubarb was used for erosion control on the banks on the canals. Now that is a good use for it. And I have to admit that last year I made a rhubarb pie. No strawberries, no apples, just rhubarb (and, of course, lots of sugar). With nothing else in it my expectations were pretty low. But it was good. In fact I might just try another one this year.

Fruit Trees- Part 1

It's late January and it's snowing. Naturally, I'm thinking about fruit trees. This is the year my orchard is really going to hit. I'll have friends over to pick all the fruit they want. We'll put up enough apple butter to last 2 years. Oh heck, I almost forgot the snow. Probably do it again in April when my apples are in full bloom. I might get skunked again.

Let's get something said right up front. Most of the Southwest isn't great fruit tree country. The single biggest hurdle we face is late frosts. You can do everything right 364 days a year but if it drops much below 32 degrees on a night when your tree is in full bloom you are sunk. Thank you for playing, try again next year. Fortunately there are a few things you can try to tip the balance in your favor.

Pick your spots. Avoid low areas, as the cold air will settle there. Keep to the ridges or the upper part of slopes. Planting on the NE corner of a house should give you some protection from the spring winds. Every year someone tells me about their peach or apricot losing blossoms to the wind. If your trees are in the open, try and stick a few evergreens on the west side to slow down the wind.

Pick your varieties. The general flowering sequence is plum-apricot-peach-pear then apple. Of course there is some overlap but almost all apricots are flowered out before the apples start. Avoid "low desert" varieties as they and will flower way too early If your neighbor has a variety that is reliable try that or something similar. (Next week I'll discuss specific varieties in more detail.)

Don't forget about the small fruits. Both blackberries and raspberries do well. Many of the old homestead orchards have blackberry patches that are still living. New berry varieties are both easy to care for and thornless.

I live in one of the coldest spots in SE Arizona. We have been 5 degrees or less at least a half dozen times this year. Because of this extreme cold air pooling, it's very hard to have productive fruit trees. A few years ago, my wife was nice enough to remind me that we've had fruit trees planted for over 10 years and we haven't harvested 10 lbs. of fruit. I thanked her for her observation. Then responded that every man deserves one bad habit and since I'd given up all my others, fruit trees were going to be mine. If only I had been able to hand her a fresh picked peach no other explanation would have been necessary. Maybe that peach would have been such a great answer that she would volunteer to help me dig holes for the new trees I plan to plant this year.

Fruit Trees- The Sequel

I biffed it. Fruit Trees was supposed to be a two-part column. I did warn you that fruit trees are my bad habit. Apples will be in their own column. For now, let's cover some fruits for SE Arizona that have proven track records. Success is measured by 1) what I've seen as the Horticultural Superintendent of the Santa Cruz County Fair, and 2) What I've eaten.

Before you choose varieties of any fruit tree you must be aware of its fertilization requirements. Will it set fruit by itself or does it need a partner? Much of the literature is confusing but almost all peaches, apricots, and apples are self-fertile. They will set on their own. Some varieties need another tree for cross- pollination. They won't set without help. Many of the pears are like this. Even if something is self-fertile it tends to do better in pairs. It is a good idea to plant more than one variety of each kind of fruit tree. Besides, it also extends the harvest season. Now for some varietal suggestions.

Santa Rosa is probably the best know plum variety and is the best place to start. They have a sweet flavor and work well for jam. Santa Rosa is self-fertile and acts as a good pollinator for many other varieties of plums. It takes 2-4 years to bear fruit.

Peaches are a highly personal thing. When ripe everyone thinks their variety is the best. And, they all could be right, as I don't think I've ever had a bad ripe peach. Nothing better than a fresh peach. The old reliable Elberta does well here in many of its forms. Hale Haven is an excellent peach that has flowers somewhat resistant to frost. Peaches can set too heavy so they should be thinned to about one peach every 6-8 inches. For you counting types, that's one peach every 38 leaves. If you have very limited space or move a lot, use dwarf peaches. Bonanza has pretty flowers, good peaches and does well in containers. Peaches take 2-4 years to bear.

Pears are the most overlooked but most dependable of the fruit trees. They tend to bear even in years of late frost. The white showy flowers are reason enough to plant them. Asian Pears (Apple Pears) also do well here. I'd start with a Bartlett. Not only does it taste good and can well, it is also a good pollinator of other varieties. Surecrop is a newer variety that flowers over a several weeks. Hopefully, some of the blossoms will avoid a late frost over that time period. Pears take 4-6 years to bear.

Cherries are almost always a bust. In the last 10 years or so I've had less than a handful of decent cherries. They seem to require lots of cold, good drainage and high humidity. We're 0 for 3 most places. If they do set the birds get a large portion of them. Cherry trees are also susceptible to root rot. It might be a good idea to get your cherries from a store. I'm not sure how long it takes them to bear because I've never seen it happen.

Heritage raspberries work well here. They bloom late, missing April frosts, and bear in August and September. Heritage is thornless and you cut it to the ground in the winter. That's easy pruning. Also, for thornless varieties try Black Satin and Arapaho blackberries. They bear on second year canes that are cut off the following winter. No staking is necessary on these varieties.

One of the current trends in fruit trees is the use of semi-dwarf rootstock. This keeps the trees smaller than they would ordinarily get. Advantages are that more trees can fit in an area, easier picking, and they bear at a younger age. Trees are more expensive in semi-dwarf form. I still prefer the standard size. Let's face it, we don't need any help keeping our trees smaller in this part of the world.

Apples - The Last Word in Fruit Trees

Apples Rule. Between eating fresh, pies, sauces, baking, and cider they are the most versatile fruit. Here are a few random thoughts to ruminate on before we get into varieties. There are more than 10,000 registered varieties of apples. This can make identifying "the little red, tart one that my aunt grew" kind of difficult. Many of our best kinds, still grown, were discovered or developed in the 1800's. Granny Smith apples grown here have the highest sugar content of any Grannies grown in the world. Must be the dry heat.

Apple trees take about 4-6 years to bear on standard rootstock. Some varieties can take longer to start. My friend, and local apple expert, Mark Douglas told me he waited 19 years for his first crop of Northern Spy (NY.- 1800) to come in! Apple trees can live 60 years or more. The information on apple fertilization in books is confusing and contradictory. The best I can figure is that with a few exceptions (Winesap and Red Delicious) most apples are self-fertile. However, they will perform better with a pollinator of another different variety. This will also extend your harvest.

Here are a few types worth trying in Southeast Arizona. Remember that apples off the tree are far better than store-bought. Supermarket apples can be picked way too early or bred for appearance instead of taste (Red Delicious).

If you're going to only put in one tree start with a Golden Delicious (discovered in West Virginia in approx. 1890). Picked early they are tart. Let them ripen and they are a sweet all-purpose apple. Golden Delicious is an excellent pollinator for almost all other apples. Gala is the yuppie cousin of the Golden Delicious. Gala was created from three quarters Golden Delicious and one quarter Cox Orange Pippin (England-1832), a dessert apple. Galas keep better than Goldens.

Jonathan (Upstate NY.- around 1800) is a small to medium red tart apple. They are good for eating, pies, and sauce. Jonathan ripens early and is often the first good apple of the year.

Granny Smith (Australia - 1868) is a local favorite and one of the last to ripen. This large, green, tart apple is many people's favorite for pies. They keep well, and if refrigerated, can last well into winter.

Red Rome (Ohio - 1840), also called Rome Beauty, is one of the last apples to flower. This can be an advantage in our spring frosts. Romes are mostly used in cooking and baking.

These are just a few of the proven varieties for our area. Of course everyone has his or her own personal favorites. New varieties are coming out all the time. A couple of "new" releases that should do well here are Jonagold and Fuji. Both are sweet and good keepers. It's hard to talk apples and not think of Johnny Appleseed. He did exist and his real name was John Chapman. He planted Summer Rambo (France - 1600's) seedlings and others. There is still one known Johnny Appleseed tree still living, barely, on one of his descendant's farm in Ohio. Let's drink a glass of cider to his work and memory next fall.

Let 'em Bee

It's fall and with the Vernal Equinox comes some of our more interesting nature viewing. Giant grasshoppers play chicken in the road, you can count dead snakes between the mile markers, and of course the bees are swarming. Drive through a swarm and for a second you might be fooled into thinking that it's raining. No such luck, but the bees are worth a closer look. Let's start with some bee basics. I should mention that in this case we are talking about honeybees, not the native or solitary bees. They deserve a column of their own.

I was working closely with the Az. Dept. of Ag. in the early '90's when the "Killer" or Africanized bees first made their appearance. It was their job to monitor bee colonies. Within a year, 80% of all wild bee colonies around Sierra Vista had been Africanized. There might be a docile European counterculture honeybee co-op somewhere in the Mule Mts., but everything else we see is probably Africanized. It is very hard to tell them apart by sight.

The swarms that we are seeing are groups of bees leaving an old colony and looking for a new home. During their search they often make a rest stop on a branch or side of a building. If they stop for very long the swarm will form a ball with the queen in the middle to keep her warm. A few bees may fly off to act as scouts to find a location for a permanent colony.

The scouts are looking for a cavity, preferably with small holes for coming and going. Hollow trees, abandoned vehicles, and parapet walls work well, especially if they are near food and water. Once a swarm sets up housekeeping and begins to produce honeycomb, they start to become aggressive. The longer they have inhabited a space the more aggressive they become. Bees foraging for food, no problem. Bees swarming and looking for a home, no problem. Bees established in a colony, problem.

If you do have a bee problem it is best to have it handled professionally. We are fortunate to have several very good bee people in the area. Check a local phone directory such as "Country Connection" for numbers. Some do live removal and relocation and some just nuke 'em. If you have to do something yourself, soapy water works as well as anything store-bought to knock them down.

Depending on your point of view, you may or may not want to attract bees with your landscape plants. Bees don't see red very well so they are not going to be attracted to red flowers. Many of the plants used to attract hummingbirds (Autumn Sage, Chuperosa, Texas Betony) don't bring in bees. Blue flowers and members of the Sunflower family attract them. Rosemary, lavender, and mints really draw them.

One of the most interesting things I learned in conversation with a local beekeeper was that a person, who is not allergic to bee venom, can withstand 6-10 stings per pound of bodyweight without it being fatal. So, I guess I'll have another plate of that good 4-H barbecue they sold at the fair, a big piece of chocolate cake and wash it down with a 64 oz. soda pop. It just might save my life someday.

Let 'Em Bee – Clarification

In the column "Let 'Em Bee" there was a statement that soapy water works as well as anything store-bought to knock down bees. Some concern was raised that this might encourage people to try and remove swarms by themselves. Bee removal should only be done by professional or trained personnel. The mood of a hive can be hard to judge. If you scatter the bees you might bring their wrath down on your unsuspecting friends or neighbors. Save your "exterminating urges" for grasshoppers.

Excuse me, Your Girdlers are Showing

Once again we are treated to the late summer sight of mesquite trees with their tips turning brown. If you take the time to look closer, you will find a perfect ring has been grooved into the base of the brown tip. That, my friends, is the work of the Mesquite Twig Girdler. And, to you entomologist types, *Oncideres rhodosticta*.

The Mesquite Twig Girdler is a narrow beetle about one and one half inches long and half an inch wide. It is mostly gray with some black and white mottling. Its long antennae, about the same length as the body, are a good ID characteristic.

Hard to believe, but it is the female of the species that cause the problem. They cut the ring around the branch with a set of powerful jaws. Then they lay eggs in the part that has been cut off. Girdling the branch insures that no sap will flow over the eggs and suffocate them. The brown tip will then fall on to the ground where they can overwinter.

Most of the time, these varmints don't do major damage. In fact, their "pruning" results in thicker, bushier trees. In the literature it says that Girdlers are almost never a problem two years in a row.Unfortunately, we seem to have Girdlers that haven't read the literature and we are in the second year of a pretty bad infestation. Several people that I've talked with have lost major portions of newly planted trees.

The not so good news is that there's no "magic bullet" for control. You should start by picking up and disposing of all fallen branches. Remember that they contain the eggs. Like many other insects, the Girdlers are attracted to lights. If you have a porch light, turn it off. Systemic insecticides don't work because the female is just cutting the bark not ingesting it.

You can spray them with an appropriate insecticide but just pulling them off and stepping on them will give you the same result and is cheaper and more satisfying. You'll have to check for them a couple of times each day as they seem active most of the time.

If you would like to view these critters by the zillions go to a Patagonia High School football game. Patagonia is surrounded by thousands of acres of mesquite trees and the stadium lights really draw them in. Last Friday night almost everyone had at least two or three of them on at one time or another. Judging by the final score, 50-8, Girdlers add to the home field advantage

The Last Word on Grasshoppers (We Wish)

It doesn't matter what the Chinese calendar says, this has been the "year of the grasshopper". Half of the folks I've spoken with groan about how bad the devastation was. The other half are grateful that we didn't get the other nine plagues, too.

The explosion of grasshoppers this year was probably the result of two warm dry winters and good early summer rains. It appeared that the grasshoppers hit the areas that got the best start on the rains earliest. By mid-August, however, we'd all been given equal opportunity to feed the swarms. The fact that they seem to be going away now is a result of the changing light cycles.

Dealing with the grasshoppers was a learning experience. The basic problem is that they migrate. You could eliminate every one in an area and two days later a new hoard would replace them. (It is kind of like our border problem.)

However, there were some techniques that seemed to lessen the damage. Here are a couple of ideas for next year. Perhaps the best place to start is to look at what got hit and what got missed. For the most part the native trees (e.g. Mesquites, Oaks, and Junipers) were left alone. The hardest hit were gardens and fruit trees. There might be a lesson there.

One of the most important steps you can take to keep grasshopper populations low in an area is to keep the grasses cut low. In the tall grasses surrounding my orchard you could see vast numbers of grasshoppers lurking, like slack-jawed youths at the food court in the mall. Where the grass was cut low the number per square yard was much less. Letting chickens, ducks and turkeys have the run of the place helps, too.

Semispore is a brand name of a protozoan (nosema locustae) that comes in grain bait form. Semispore is very host specific and attacks the digestive system of only grasshoppers and crickets. It has no effect on anything else that consumes it, or eats the grasshoppers that die from it. Also, it is inexpensive and easy to apply. Unfortunately it must be applied early, when the first grasshoppers are 1/2"-3/4" long to be most effective. The good news is that Semispore also affects grasshopper eggs too.

DeBug is another grain bait. It contains a mild dose of Sevin. Grasshoppers eat it, and they die. It's a beautiful relationship. It has low toxicity to other animals (non-bugs). To be effective, it needs to be applied regularly.

The most effective solution I came across was the application of a systemic rose food to the plants that I wanted to protect. A systemic food kills the insects that eat the plant, while feeding the plant at the same time. Often insects will avoid a plant that has been treated this way. Although effective, this method has some downsides to it. It's expensive to treat large plants or large numbers of plants this way. And, I certainly wouldn't treat fruit trees with fruit on them with a systemic as it would poison the fruit.

Our best hope for lower grasshopper populations next year is definitely a wet winter. When you see them, act early and often. If all else fails, maybe we can find a good recipe for grasshopper burgers.

Meanwhile, Back in the Orchard...

Let's add another pest to the list of things we have to deal with in the home orchard. In the past 4 or 5 years I've had problems with deer, gophers, grasshoppers, rabbits, and porcupines. Now I have the joy of daily checking for Sapsucker damage. If you have noticed lots of small holes in a circle drilled around a branch or trunk, think Sapsuckers, Flickers, or Woodpeckers.

The holes made by these woodpecker types are usually about ¼" in diameter. They come close but don't quite girdle the tree. One might think that they are going after insects that are burrowing just under the bark on your trees. One would be wrong. The holes are drilled in order to make the sap run. The sticky sap attracts and catches bugs. Periodically the birds stop by to see what is for lunch. It's just like the old fur trappers.

The good news, according to Rob Call, Cochise County Extension Agent, is that he has never seen a tree die because of this kind of damage. If left alone, the tree will heal on its own. It should take a couple of years for the holes to close completely. Do not use pruning paint or tar. Studies have shown that not only does it not help, but it may even slow down the healing process. This goes for all plants not just fruit trees.

The best way to deal with pecking bird damage is to practice exclusion. Exclusion is county agent talk for putting some wire around your branches to keep them out. I've been using small mesh chicken wire and metal lath. As soon as there are a couple of holes I wrap the branch or trunk, extending as far as I can above and below the damage

Being a normal, thoughtful guy I thought the best solution was just to shoot the offending birds. Fortunately, before I put my plan into action I was informed that it was both unwise and illegal. Almost all bird species, except game birds, are protected under Arizona law. Don't shoot them.

While wrapping trees with metal lath (my fingers were bleeding from being repeatedly poked by sharp metal) I was thinking about all the critters that have plagued the orchard. Add to the mix late frosts, drought, and other assorted climatic conditions. I had to ask myself was it really worth it? Then I thought back to last year's crop. 7 Rome apples, of which some unknown varmint stole, and 3 peaches. Was it worth it? Heck no! But, maybe next year.

Arizona Cypress Borer

(Jim's Note- Warning: this article contains little or no entertainment value. It should be read by only those with problems with their Arizona Cypress or those having trouble sleeping. It is boring.)

Every fall there are lots of questions about the brown tips on Arizona Cypress trees. Usually the last 4"-6" have died and are hanging down at an unnatural angle. In some extreme cases the top third of the tree has browned out. The worst case scenario has the whole tree dying. Of course the question is: what causes this and what should be done about it?

The easy part is the cause. These Arizona Cypress trees are being affected by the Arizona Cypress Bark Beetle. This insect isn't real particular and can bother our native junipers and a few other ornamentals as well. Three or four years ago it hit most of the Alligator Junipers in the canyon here. In fact, starting in 2003 there has been a huge jump in the amount of damage done state wide caused by the various kinds of bark beetles. It is estimated that over 3 million pine, juniper, spruce, and cypress trees have already been killed in Arizona.

The culprit here is only about the size of a pencil lead. The ones that I have seen are a brownish-black color. They bore into the lateral branches of your Arizona Cypress tree and lay their eggs. When the eggs hatch they eat the soft tissue on the branches which, unfortunately, is the part that carries food and water to the tips. With no nutrition the tips die. If you look closely at the base of the dead tip you will see a small hollowed out place where the eggs were laid. If the tips are already infested the beetle will move on to the trunk where it can also bore in and lays its eggs. When the baby beetles hatch they eat under the bark and create tunnels which look like some kind of primitive writing. Again, if there is enough tunneling, the trunk can be girdled and the tree will die.

These beetles are built to survive. They prefer stressed trees and will attack them before getting started on healthy ones. A healthy tree produces plenty of sap that physically repels the beetles as they try and bore in. A stressed tree can't produce enough sap to keep them out. The bark beetles have always existed in our forests; it is just that now the drought prevents the trees from fighting back. (You knew I couldn't go a whole article without blaming something on the drought.)

I am assuming that if you are reading this after the warning at the top of the article, you are probably having problems with your tree and are wondering what to do about it. First of all, check a couple of symptoms to make sure your problem is caused by the bark beetle. If you have dying tips and/or sap flowing from holes the size of a pin bored in the trunk, you probably have Arizona Cypress Bark Beetle. If you can suddenly see parts of your neighbor's '66 Chevy that he is rebuilding but which used to be hidden by the trees, you are probably wondering what can be done to save them. I am here to tell you that if your tree is already infested, and has severe die-back, there are only two options. Learn to appreciate the craftsmanship that went into the legendary Chevy 250 straight-six engine, or move.

If your trees are showing only minor injury like tip damage, there are some preventative measures that can be taken. Water deeply during the growing season. A healthy tree is the best defense. You can try and remove some of the damaged area by pruning. Get rid of any material that you have cut off or dead trees that have been cut down.

They could hold eggs waiting to hatch. The use of insecticidal sprays has proved absolutely useless. Like any other insect the population of the Arizona Cypress Bark Beetle grows and shrinks. A couple of wet years and it won't be such a problem anymore. Then you can talk your neighbor into planting a few new trees.

A Sad Day in Santa Cruz County

It is a sad day in Santa Cruz County. For a change the problem isn't lack of moisture, grasshoppers, or late frosts. I am referring to the new policy at the Sonoita Landfill.

About a month ago I was making my monthly (well, quarterly) dump run when I saw the new sign. It basically said "no salvaging or scavenging at this facility". The sign was just off to the right of the recyclables bin. My first reaction was that this is just plain stupid. Later the irony set in. We can bring items to the dump, pay to have them hauled off, use the energy to recycle them, and then buy them again. But taking a 2x4 home (which is now illegal) and using it again is not "recycling", it is scavenging. Please remember that we are also trying to save trees, and landfill space is at a premium.

Several years ago Jane Woods told me that local folks have been recycling here long before the word "recycling" became a part of the vocabulary. She told me that if you had to take something to the dump that was still good, it would be put to the side where some lucky person would claim it. I am sure it made the person who hauled it in feel better that their treasures had found a good home. The nicest thing on my courtyard is an old time galvanized stock tank that I salvaged from the dump and now use as a fish tank. It only leaked a little bit at first. All of my gates have some metal in them that I got from the dump. Whenever I had to build a new one I made a dump run. We called it shopping. In fact before I saw that hideous sign, I didn't know that you could buy steel new.

Going to the dump used to be fun. Everyone who went enjoyed Angel M.'s, the friendliest guy at the landfill, creations. His arrangements of discarded stuff were a lot more creative and entertaining than most of the work that comes from so-called trained and educated landscape architects. If you are new to the area, sorry you missed it because those were salvaged materials he was working with.

Dealing with a moronic bureaucratic policy is trying but I still feel the need to attempt to be fair. Could there be a legitimate reason for the change? Probably not, but if I was to ask I am sure the answer would be the refrain that has taken so much fun (and personal responsibility) out of our American life, "but, what if?" This is used as in "But what if someone took a piece of metal home and cut themselves on it and got blood poisoning? They would sue us". The "but, what if" policy usually only applies to a small percent of cases but ruins it for everyone. But you know what? I think if I cut myself it is my own dang fault. I'll deal with it.

I can't stand it when people gripe about a problem and then don't try and find a solution, so here is mine. This is a county election year. My vote for supervisor is up for sale. The first candidate that publicly states that he will allow salvaging in county landfills gets my vote. I don't care if they are Republican, Democrat, Independent, or Goat Lover. I am voting for them. And hopefully this winter I will be fixing my chicken coop roof with the salvaged campaign signs of the losers.

Dump Redux

There are two things I'd like to cover before we get back into our discussion about our local landfills banning salvaging. First of all thanks to all who commented on the first column. That particular column ("A Sad Day in Santa Cruz County") generated more interest and response than anything else I have written with the possible exception of the column about the poisoning of the native sunflowers on the highways. If I am ever feeling like I need to raise the column's ratings I will make up a story about poisoning the sunflowers at the dump.

Secondly, I really have absolutely no clue what "redux" means. I know that it was part of a title of a famous book and feel that it gives some much needed sophistication to this topic. In other words, I liked the way it sounded.

The last time we got into the topic of scavenging being banned at the landfills, I announced that my vote was for sale to whichever candidate would change that policy. One candidate did take the time to call and explain his position. John Maynard explained that although he personally thought salvaging was a good idea, the law banning it was a state statute and not a county one. He even read me the A.R.S. number which I failed to write down. I also talked with an employee of the landfill who told me that he also hated to see good material thrown away and buried but there were legitimate safety issues with the dozer. It is hard to operate it safely if people were scurrying around looking for the good stuff.

These are both legitimate reasons to explain the current policy but they are not insurmountable obstacles.

First, dozer safety. We really don't want the good citizens of Santa Cruz County to become part of the permanent collection in the landfill. That means we have to keep them out of the way of the equipment operators. The easiest and best way to do this would be to create a separate area in the flats before you enter the pit where the good reusable material could be off loaded and salvaged. This would be good for lumber, steel, and other building materials. Every so often a loader could come by and pick up material that had gone unclaimed for say a month or so, and put it in the pit. This is basically what is being done with the other recyclables (plastic, glass, etc) except that we have to haul that stuff off site.

As for the state statute that bans salvaging, well it is wrong and needs to be changed. I hope that our candidates for supervisor and state representative can see that. My vote is still for sale.

We Lost A Good One

The cactus world lost one of its best friends last month when Dave Eppele passed on unexpectedly. Most of us knew him as the founder and driving force behind Arizona Cactus and Succulent Research Inc., otherwise know as "the cactus place in Bisbee". If you did know him through Arizona Cactus you probably called him El Jefe, which was how he signed his syndicated columns that appeared in various newspapers in the Southwest. I always called him Tio (Uncle) out of respect for what he had done and what I learned from him.

Dave was a bonified Western character. He grew up in Northern New Mexico. Many of his best stories were about growing up in that area 60 years ago. Remember the time he told about eating turkey cooked in red chile with the Navajo family at Thanksgiving? How about the time his father, a deputy sheriff, brought in a prisoner using the leaves of a desert spoon as handcuffs? Dave didn't live in the past, though. He was passionate about current western issues, especially water usage. Above all Dave was an educator.

Arizona Cactus was Dave's baby. He created a botanical garden to display and see what cactus and succulents would work in the high desert. Admission to the garden was free. "Something in the West still has to be free" Dave would say in his gravelly voice. He would lead tours himself. He didn't care if it was one person or fifty. You got the same treatment if it was tourists from Minnesota, who didn't know a prickly pear from a saguaro, or a PhD university botanist. Actually I think he preferred the common folks as they were easier to reach about the importance of appreciating and respecting our arid land. The one thing Dave couldn't stomach was botanical arrogance. Try and insist on a correct pronunciation of Latin names (a popular pastime of cactus fanatics) and Dave would be all over you like a starving junk yard dog. It was important to him to always keep it real. The highlight of the year at Arizona Cactus had to be the Labor Day Fiesta. El Jefe and a small army of volunteers would go all out. The food was outstanding. There was cactus and eggs, cactus casserole, vegetarian cactus casserole, cactus and beans, cactus salad and cactus tea. If Dave knew you, there might be a shot of cactus wine. About the only thing that didn't have cactus in it was the fine red chile sauce he made. My band, The Busted Cowboys, has played there for years and no one ever made us feel more welcome than Dave did.

Dave did things that the rest of us can only dream about. Most of his cactus collecting days were before Mexico changed its laws of plant collecting. He went places and did things that just can't be done anymore. Don't think of El Jefe as just a plant guy, though. After all, he still had to make a living. He didn't talk much about his other "interests" unless he really knew you. Dave was an extremely talented musician. At one time he toured with Duke Ellington and his orchestra. New Mexico State gave him a musical scholarship. However he gave that up to pursue a different path, playing "houses of ill repute" in Juarez. At one time Dave was the radio voice of the old Denver Broncos when they were still in the AFL. And if that isn't enough, he had the first talk radio show in Albuquerque long before talk radio became popular.

And so, Tio, I think somehow you are perched on a giant saguaro looking down on all of us left doing your work. Trying to get people to understand that water is precious and nothing beats a good chile. I'll keep imitating your distinctive voice when I describe a plant as needing "no supplemental irrigation". I'll use scientific names only when it helps to make a discussion of a plant clearer, not just to sound good. I'll also try and get people to understand that most of SE Arizona is Chihuahuan, and not Sonoran, Desert. I will keep stealing your line about plants growing where they aren't supposed to according to the maps because "plants can't read". Somehow though, the Southwest just seems a little blander, a little more vanilla without you. Maybe we need another shot of that cactus wine.

We Just Flew In From Mexico and Boy, Our Wings Are Tired

Those of us fortunate enough to live in Southeast Arizona live smack dab in the heart of a migration zone. No, I'm not talking about the RV's heading to the parking lot at the Wal-Mart in the winter. We are in the migration path of hummingbirds and lots of them. (Please note that I refrained from making a comment on illegal immigration as I thought that would be way too obvious.)

This part of the state has about 14 species of hummingbirds. Twice a year they move through here. In the spring they are traveling from Mexico to, for some species, as far as Alaska. In late summer and fall they reverse direction. Right now we are in the peak of this movement. It started in July and should be over in October.

Different species of hummers migrate at distinct times. Of course there is some overlap. Right now we are seeing a lot of Black-Chinned. In September the Rufus Hummingbirds will peak. October is good for Anna's.

Obviously this great journey takes a lot of fuel. The little birds are big eaters. Protein comes from small insects and they devour plenty of them. Most of the food that mothers feed their young is insects. In addition to the protein they also need a bunch of carbohydrates. That is where we can help out.

Most people try and attract hummingbirds with feeders. Apart from being kind of messy and boring it works pretty well. The bees and nectar feeding bats really appreciate your efforts as well. If you are wondering why your feeders are being drained at night it is probably the bats. Bats need love, too.

The best way to attract and feed hummingbirds is with your plants. When choosing plants keep a couple of things in mind. They should be nectar producing and they need to be in bloom during the migrations. Many native or southwestern plants work well for this purpose. Look for those with tubular or trumpet- shaped flowers. Orange or red are good colors but not always necessary. Plants that have their flowers at least 18" off the ground are a good choice. Otherwise your friendly neighborhood kitty might try and make the hummers into Kentucky Fried Hummingbird.

In slightly lower elevations, ocotillo is a very important natural source of food. Most of the penstemons work well, too. Scarlet Bugler (Penstemon barbatus) is native right here. It has tubular orange-red flowers and will tolerate some shade. Anisacanthus thurberi is also known as Chuperosa, which translates as hummingbird. It grows in our washes and blooms in the spring. Desert Willows with their bell shaped lavender flowers also do a good job as an attractant.

For the fall migration you could start with Chuperosa's cousin from Texas called Mexican Flame (Anaisacanthus quadrifidus). It waits till the heat of the summer to start to bloom. Then it is covered with red-orange blooms. Texas Betony, also known as Scarlet Hedge Nettle, is another shade loving plant in bloom now with coral flowers.

A few plants are in bloom for both migrations. Tops on that list is Autumn Sage, which has as long a blooming period as any plant does here. Red Yucca is an easy one to care for that also blooms for a long time. Coral Honeysuckle blooms in the spring and the fall and likes an eastern exposure.

The plants mentioned above are just a few suggestions. There are lots more. If you plant it they will find it. It is kind of nice to get a second return on your work and watering. One time for the flowers and a second time for the hummingbirds. I really need to thank local hummingbird researcher and expert Susan Wethington for her help. Susan supplied all the real information for this column. The rest I just filled in.

A Fair Project

We are getting close to my favorite event in all of SE Arizona, The Santa Cruz County Fair. The Fair, more than any other local event, reflects the true character and traditions of the people that live here. The success of the Fair depends more heavily on local participation than any of the other big events that are held at the Fairgrounds. Being an active participant in the Fair is a chance to learn, meet some new people, and get reacquainted with old friends. Getting involved in this particular event is easier than you think.

Yes, many of us were involved in the horse races in the spring. Sometimes the involvement was the kind where you sit in the stands and donate money every 15 minutes or so. Of course, if you study the program or have a system, once in a while money is donated back to you. I have a system. I bet the jockeys. I figure that the best trainers will hire the best jockeys for their best horses. My system usually loses.

The Rodeo is another event where most of us choose to participate from the stands. It is coming up over the Labor Day weekend and we should all support it. I hadn't been to the Rodeo for a couple of years so I went last year. Things sure have changed. I hate to admit it but I was a little disappointed in some of it. The roping was fun to watch but the "crowd favorite", bull riding, turned me off. I know that the days when a local rodeo could afford to hire live music are long gone, but this was ridiculous. During the bull rides they blasted two kinds of so-called music. What was wrong with George Strait? The first and more pleasant to listen to was heavy metal. For those unfamiliar with this style it basically sounds like a large internal combustion engine running out of oil at high speed. This was better than the second that was played, which was rap. I have always heard that the sport of rodeo was an organized way of cowboys demonstrating the unique skills they used in everyday life. What the heck does rap, an inner city urban music style, have to do with that? Maybe a compromise can be worked out here. No Hank Williams or Patsy Cline at your drive-bys or break dance contests, and no rap at our rodeo.

Ah, sorry folks. I may have gotten a little off course here. Let's get back to the County Fair and how to get involved with it. The first thing you can do is to become a member of the Santa Cruz County Fair and Rodeo Association. Although it is call the Santa Cruz County Fair we are not a government organization. This is a private group that anyone is welcome to join. Just come down and give Tina a little bit of money and she will sign you up. Your membership card gives you voting privileges and free entry to most events on the grounds.

The Fair itself is a combination of many different interesting areas. Horticulture, floriculture, photography, cowboy crafts, schools, livestock and the arts all have their own exhibits. You can learn more about what grows well in your area by helping to take horticultural entries than can by reading any books or columns. Do you like cooking or quilting? Come down and help with that department. If you don't have a specific area you want to work in, you can help out by watching the admission booth. There is always help needed with that. Just pick up a Fair Book, available at the fairgrounds and elsewhere, to see who is in charge of a department or call the fairgrounds (520-455-5553) to volunteer. Who knows, you might even be a superintendent next year.

One more thing. We are planning to plant a tree in honor and remembrance of Richard Harris, our former County Extension Agent and a good friend of the Santa Cruz County Fair. He was always someone we could count on to help us out this time of year. We will be thinking of him and his great family when the horticulture is being judged.

Back Packin'

Last fall I had the opportunity to train to become a "Grasslands Range Monitor." Grasslands Range Monitors visit various pre-selected sites on local ranches in order to identify the vegetation, mostly grasses, and record how often they occur. This information can be used to assess the health and trends in range conditions. I look at it as getting paid to spend some time with friends, walk around, and look at plants. Good deal, right? It is, but as one of my occupational heroes Paul Harvey would say, "Now here is the rest of the story."

Any job requires its own set of tools. In this case we need cameras, writing utensils, tape measures, ID books, clipboards, monitoring forms, a magnifying glass, foldable rulers, a GPS, water, a compass, and a square frame. Many of the sites we look at are not accessible by truck, so we hike in. This is where a backpack comes in handy. When I showed up for work, I borrowed my wife's college backpack. It was the one she used at the University of Arizona back in nineteen – well, it really doesn't matter when she used it. What matters is that while it may have been state of the art back then, it only has two pockets. It seems that backpacks have undergone almost as much change as telephones. Everyone else's backpack had at least 20 compartments so that each item carried had its own pocket, and they didn't have to dump everything out on the ground to find a pencil. I will admit it. I was soon suffering from "backpack envy".

About a week later I was working on a ranch in the Elgin area. While pawing through the contents of my two-section backpack, I mentioned to my friends Rukin and Joe that on my first open day I was going to the city to buy a new pack with lots of pockets. I think I remember Joe saying something about "don't bother." A week later a miracle happened. Five almost brand-new backpacks showed up at my nursery. I felt like the King of Monitoring. The only problem was picking the best one for the job.

Two of the large gray backpacks had netting between the upper and lower main compartments. Fortunately, my oldest son was around to tell me that the netting was used to carry around your skate board. Most of the monitoring sites here are too rocky to skateboard, even if I did have one, so the choice was down to three. One was canary yellow and I don't like to wear cloths that are brighter than me, so we are down to two choices. The red and black one was made by Nike and felt pretty comfortable. Now I know that you are thinking "How do I know it is really made by Nike and isn't an imitation?" Folks, counterfeiting name brand products is illegal. The people that left the back packs in this country wouldn't break the law, would they? Any rate, the last backpack had the most pockets so that was the one I chose.

So how did all these backpacks get here? That is a simple one. They are left here by people that cross the border to smuggle drugs into this country. Once they drop their load of drugs, they change their clothes (I am guessing so that there is no drug residue on them) to a fresh set which have been carried in these backpacks. Then they abandon the old clothes and backpacks. At this point they go up to the highways, hopefully get picked up by the Border Patrol, and get a free ride back to Mexico. Is this a great country, or what?

(Jim's note: If you are new to the area or unfamiliar with border life you might want to reread this paragraph.)

I went through the contents of the abandoned backpacks and found them kind of revealing. There was an airline boarding pass to Mexico, lots of toothpaste, toothbrushes, Alka Seltzer, antacid tablets, Pepto Bismol, a Mexican version of Maalox, and white and pink toilet paper. When taken as a whole, you would have to say that these smugglers are as concerned with the passage as they are with the journey.

Packin' It In

A couple of weeks ago we talked about Grassland Range Monitoring which led to a discussion of the amount of backpacks that have been left in the hills and canyons. Folks, the backpacks are just the smallest fraction of the amount of crap that has been dumped in our country. There is not a ranch near the border that hasn't been used as a garbage dump by illegals entering into the Southwest. This is situation where if you haven't seen it for yourself, you can't understand how bad it is. Often mounds of backpacks, clothing, energy drinks, candy wrappers, sardine cans, and used toilet paper are left at an abandoned campsite. And where are the largest of these dumps? Of course next to springs and the few permanent water holes up in the mountains. These are the same areas that large amounts of money are spent on trying to protect them by excluding them from cattle. On one ranch I visited there was an area that had been fenced to keep cows away from a stream that contained an endangered fish. That area is now heavily used as an illegal camp. I kind of doubt they are too worried about the status of the fish.

I realize at this point you might be wondering why the obsession with a little litter when there are lots more serious issues linked to illegal immigration like people dying, strain on our medical services, and fires. The answer is two-fold. First of all, in this column, we try and keep our discussion basically on plants, landscaping and land-related issues. This is definitely a land-related issue. Secondly, until very recently, there has been no condemnation by the so-called environmentalists. Do they think that if a hunter or hiker drops a candy wrapper that is littering ("pack it in/pack it out") but if a non-American leaves used toilet paper under a Manzanita that is a "sharing of culture?" The real problem is that much of what passes as environmental or ecological concerns has been derailed into a political movement, not a scientific one. Words like "ecology" have lost their true meaning (the science that deals with the relations between all living things and the conditions that surround them) and have become code words for politicians and advertisers. Not that there is much of a difference between the two groups.

If you are told someone "cares for the environment" you probably get an image of a person driving a hybrid vehicle on their way to a recycling center, not of a rancher building fence to protect a riparian area. The fact is that many local ranches here have done excellent jobs of protecting and improving their land, or environment, through good stewardship. I realize that the evolution of the meaning of words happens all the time. The word "dude" certainly has changed in recent times. And I am pretty sure that when Rex Allen sang that "was once in the saddle I used to go dashing, was once in the saddle I used to go gay" in the great old ballad "Streets of Laredo" the meaning might have changed. What we need is a word that isn't politically charged and that describes someone that makes the effort to show respect and good stewardship of the land whether it is their own or public. The word would describe someone who takes the time to understand how all the pieces fit together and that there is a relationship between soil, water, plants, and animals. I recommend that we leave the new politicized words behind and go with an old and respected term "conservationist." Trashing up our mountains is not a human rights, nor a jobs, issue. It is just flat out wrong. Any conservationist would know that.

WINTER

TRUE CONFESSION   p. 74

ESKIMOS GOT NOTHING ON US   p. 75

TALKING DIRTY   p. 76

WINTER WARNINGS   p. 77

CUT IT OUT   p. 80

"LIVING" CHRISTMAS TREES   p. 82

THE BIG MONEY   p. 84

THE REALLY BIG MONEY IN PINUS   p. 86

CATALOG COMPLEX   p. 88

TREE TOMATO, THE REST OF THE STORY   p. 89

SUDDEN OAK DEATH   p. 90

SUDDEN OAK DEATH CONTINUED   p. 92

AND THE WINNER IS...   p. 93

PLANTS OF THE YEAR 2003   p. 95

PLANTS OF THE YEAR 2003 CONTINUED   p. 97

PLANTS OF THE YEAR 2004   p. 98

PLANTS OF THE YEAR 2005   p. 100

HOW TO PLANT A ROCK   p. 103

THE CONFERENCE REVIEW   p. 105

THE RETURN OF FIVE EASY QUESTIONS   p. 107

True Confession

Time to use this most public of forums to make an admission. I'll come right to the point. I am a weather-aholic. Most of the time, I engage in this behavior in the privacy of my own home. I try to keep it from affecting my family or business. This week, however, I have to let it out, as it is the anniversary of one of the greatest weather events within the last twenty-five years, "The Great Freeze of '78".

December 7, 8 and 9 of 1978 were three of the coldest days ever recorded in Southeastern Arizona. Official low temperatures were -4, -2 and 5 degrees. The daytime temperatures never warmed up enough to thaw things out. The all-time official recorded temperature for this area is -6 in 1949. Some people have told me that in Elgin their temperatures were down as low as the -20's.

The low temperatures tell only half the story. The interesting part is that up until this time it had been a mild winter. Plants were not hardened off to the cold. The abrupt extreme change caused damage even to native plants such as the mesquite and junipers. You can imagine what it did to the local plumbing.

At the time, I was a young, perspiring - I mean, aspiring, landscaper in Tucson. Down by the Rillito River it reached 13 degrees. The official low was 20. Trees that never had a problem, such as the Aleppo Pine were severely set back. On December 6th many old places had cactus fences that were 10-15 feet tall. On December 7th, they were turned into piles of green mush. When talking with my friends about this freeze we still refer to it as "The Great Cactus Massacre".

The real question is, are we due for another great winter weather event? It's been over twenty years since the big freeze. Fourteen years since the last big snowstorm. On the other hand, the official high for the area is only 109 and that goes all the way back to 1926. Mark my words, something is going to happen.

Eskimos Got Nothing on Us

Common lore has it that the Eskimos have over 18 different words to describe snow. Anthropologists say this demonstrates the importance of snow to their culture. Should any anthropologists spend a summer in SE Arizona they might reach a similar conclusion about rain here. We have quite a few terms to describe what falls from the sky. To make their life easier I have decided to put together a glossary of rain related phrases. I did not make any of them up and have heard or used them all in common conversation.

GLOSSARY

(Warning some of the terms may appear quite graphic. The forces of nature are not always polite. Do not read them if you are easily offended.)

8 inch rain – This phrase is used humorously to describe a very sparse rain. "We had an 8 inch rain – 8 inches between the drops."

4 inch rain – This is actually better than an 8 inch rain because the drops are closer together.

Dust settler – Just enough to cover the ground. If there hasn't been any rain for a while, a dust settler might get your hopes up.

Winter rain – A slow steady rain over a long period of time, usually at least a couple of hours. A winter rain can happen any time of year.

Gully washer – A good amount of rain that comes down quickly. More falls than can soak in so the draws, canyons and low spots become temporary rivers. The first gully washer of the year helps to clean out brush and debris that have built up over time. This process used to be an accepted way of cleaning up your property.

Frog strangler – Lots of rain in a short time span. The idea is that so much moisture comes down so fast that it could drown amphibians.

Toad choker - See frog strangler.

Tank filler – A rain or series of rains that puts water in cattle tanks. A couple of good tank fillers, and ranchers might not have to haul water for a while. (Hauling water doesn't pay real well.) Tank fillers can happen with a sudden heavy rain or by any decent rain if the ground is already saturated.

Turd floater – This is the sweetest rain of all. So much rain falls that the ground and everything resting on it gets completely saturated. The manure (and anything else) that was drying up in the fields is lifted from its place of deposit and moved to a lower elevation. This is life affirming proof that it does flow downhill.

I hope this glossary helps out. One more thing needs to be brought up to understand local rain vernacular. Many people measure moisture in 100ths. They don't get a half inch they got 50 one hundredths. It kind of makes me wonder, how many hundredths of snow would it take to move some caribou poop?

Talking Dirty

The Yule log, stockings, fire in the fireplace, tis the season to talk dirty. Several people talked dirty with me last week and I enjoyed it. They asked," should you add your fireplace ashes to the dirt when planting? That's the way we did it where we came from. The answer is two words, definitely not.

Our soil here in southeast Arizona is pretty typical of soils in most of the arid west. High alkalinity-low fertility. The ph is high and there is very little organic material in the dirt. Ph refers to the acid — alkaline balance in the soil. Low numbers, between 0 and 7, are acid and higher ones 7 and above, are alkaline. 7 is neutral. The ph scale is like the Richter scale in that material that tests out at a ph of 8.5 is ten times more alkaline than something at 7.5. Fireplace ashes are very alkaline. Adding them to your planting mix would raise the ph and limit the types of plants that would be happy growing in those conditions.

In the east and areas of high rainfall (over 20") much of the soil is acidic. This is a result of the higher rainfall promoting more plant growth. When the plants die they decompose in the ground and raise the amount of organic matter. The ph is lowered. Using ash there might help neutralize soil conditions in a very local area.

One of the characteristics of our alkaline soil is that it tends to bind up nutrients. The two most common problems that we see are nitrogen deficiency (yellow leaves on older growth) and iron chlorosis (green veins on yellow leaves of the new growth). Many times these nutrients are present but unavailable due to soil chemistry.

Should you spend lots of money at your local nursery on peat moss, mulch, and soil sulfur to improve your soil? No, that is very expensive and doesn't work over the long haul.

Unless you want to do it on a constant basis you are just pouring money down a rat hole. It's better to avoid acid loving plants such as azaleas, gardenias, and blueberries. If you feel you have to have them put them in raised beds or planters. The best solution is to use the plants that thrive in our conditions with no special care. Go make soap with your ashes.

Winter Warnings

The past few weeks have been absolutely beautiful here in Southeast Arizona. Days in the mid 70's, and cool nights. Sure makes for good sleeping weather. There was an early cold spell in October but that didn't last long. It doesn't get much better than this, does it? Even the most skeptical of us can't see a cloud on the horizon, can we? Of course we can!

First of all, it is dry again so the fact that there are no clouds on the horizon is <u>not</u> a good thing. However, for the time being, we aren't (believe it or not) going to discuss the lack of moisture. There are a couple of other related weather points that we should be aware of.

Have you looked at your plants lately? Most of them have stopped flowering but don't look much different than they did 6 or 8 weeks ago. The warm temps have kept them going when they should be going dormant. I looked at some container plants from Tucson the other day and they had new growth on them. An Arizona Walnut that had dropped its leaves had new white roots developing when it came out of the bucket. "That's good" you might say. No, that is bad. Plants need a time to harden off to the cold. This year is starting to remind me of two years ago (2003) and that legendary year, 1978.

In both those years it was a very mild fall. It stayed relatively warm up until December. Then the cold roared in with a vengeance. Sap was still flowing in many plants. Plants like peaches, Butterfly Bushes, even some pines that should have no problem with the cold were stopped dead (literally) in their tracks because they hadn't hardened off.

So far this year has the potential to be like those years. If that does happen you will probably be more concerned about your pipes than your plants, but you could avoid some problems by covering them with sheets or blankets.

And here is something else to start to think about on this beautiful fall day. We are overdue for a Big Winter Storm. We haven't had one for a while. The last decent snow to hit was in the winter of '87. I am not sure exactly what came down but I do remember that it fell on a Tuesday and on the following Saturday there was still 6 inches on the north side of the roof. How much snow would it take to be a Big Winter Storm? That is hard to say because of a few reasons.

There are three legitimate ways that weather events like a snowstorm can be recorded. The first and most recognized is by use of official recording stations. These have actual scientific instruments and tell you what the conditions were at a specific site. I like to use the information from the Canelo Station because it has almost 100 years of records and hasn't been affected by urban sprawl like weather stations in big cities like Tucson, Benson, or Sierra Vista.

The problem with the official stations is that they can only give you information from one specific site. As we all know, especially for moisture, what happens a half mile away is often different that what you got at your place. The second method for recording weather is by asking "what did you get?" Find out what enough people got and you will have a better picture of what really happened. A good example of this was last summer's rainfall. Most people got somewhere between 8" (average) and 12" (excellent). This tells you that it was overall a decent year for summer moisture.

Temperatures are another thing that it is good to get local info on. The official low for this area is -7 degrees in 1978. However folks that were living in Elgin at that time claimed to have dropped down to the -20's. That is a big difference.

The last way to gauge weather is by recalling weather memories. This is by far the most fun. We all know that it used to be hotter and colder, wetter and dryer, and windier than it is now. My old friend Wayne Wright told me that he heard about a storm in the teens that put 48" of snow on the ground near Greaterville. The rancher who told him the story said he ran out of tobacco and coffee so he saddled up two horses to ride to town (Sonoita). He would ride one until it played out and then get on the other for awhile. Wayne also said that 16" snows were common until recently and that you always measured them on fence posts so the measurements were accurate and not from drifts. Now doesn't that tell you more about the weather conditions than the barometric pressure and dew point numbers?

Now you are asking "what does all this have to do with this being the year of a Big Winter Storm? And, you said people claim it used be windier, well this article is getting pretty windy too; how about some real information?"

Ok, here are some real boiled-down numbers for you to think about. The largest snowstorm ever recorded here was 23" in 1916. Average snowfall in Canelo is 5.91". In Elgin it is 5.6". Patagonia is about the same. The last real Big Winter Storm was in 1978 (that date keeps coming back, doesn't it?) when 11" hit the ground. The year that had the most snow was 1919 with a total accumulation of 39." If you like the cold white stuff you should have been here from about 1911 – 1921. For those 10 years the average snowfall was 18.84". For comparison, the winters of 1994-2004 averaged just less than a ¼".

Weather memories bear out the recent snow scarcity, too. When my older kids went to school, every year they had at least one snow day off and usually several of them. I will never forget leaving Elgin School - I think it was the first year the new school was open - on the night of the Christmas Play. The snow was coming down in a blizzard. You could barely see the lines on the road. These days kids here don't know what a snow day is. There hasn't been one in several years. That might change this year. I am telling you we are overdue.

Cut It Out

For those of you so inclined, there is still some time to do some pruning. Most plants can be cut on anytime but pruning when they are dormant is probably best. Before you break out the old Coronas (a popular brand of pruning snips) you might want to ask yourself why are you doing this? Entire books have been written on this subject, but I think we can cover much of what is important in this column with a few questions.

Q. Can I plant something under my window that grows to 6' if I cut it back a couple of times a year?
A. It's your house, you can do whatever you want. But, why not plant something that grows to 4'? Then you don't ever have to deal with it. If you have to constantly cut on a plant to keep it contained, rip it out or move it and replace it with one that requires no additional work.

Q. How do you prune Texas Rangers (or any plant) to make those perfect geometric shapes?

A. You don't. People that want plants in those shapes need more help than can be given here. Animal shapes are pretty neat, though.

Q. What is the proper way to prune fruit trees?
A. Lightly. Start by removing dead wood and branches that cross back into the tree. Gently try and give the tree some shape but don't overdo it. Mark Douglas, a local friend who has probably produced more fruit than all of the rest of us combined, hardly prunes at all. He feels that leaving the trees thick gives them some self-protection from the sun's harsh UV rays. Much of the pruning that goes on in a commercial orchard is for ease of picking or access for equipment.

Q. Any other reasons for keeping them woolly?
A. It makes for a healthier plant. Research shows that it is the chemicals (auxins) produced in new leaves that stimulate much of the new root growth. The saying among serious tree people is "prune a little tree a little". My saying is "it is easier to cut it off later than grow it back now".

Q. Speaking about "serious tree people", who can I call if I need real help?
A. If you are dealing with large trees, trying to bring back long neglected fruit trees, or wondering about the health of old trees, call a certified arborculturist. These are people that have taken extensive training in accessing and working with trees. They tend to be well paid for their services.

Q. When should I cut back plants that have died back over the winter?
A. Cut them back after the new growth has appeared and you can tell what is actually dead. If you are not sure how much is dead, scratch a branch with your thumbnail. If there is green showing, that part is still alive. If not, go lower on the branch and try again.

Q. Should I always use pruning paint or tar?
A. Never use that stuff. Research show that plants will heal just fine on their own. Pruning paint is just a waste of money.

Q. Where can I find some "real" information about pruning that is not so opinionated?
A. The best condensed, common sense, straightforward guide to pruning that I've come across is in the 2003 "The Almanac for Farmers and City Folks". It's on page 100 if you are planning to read it at the grocery store.

"Living" Christmas Trees

Today we try to unlock one of the world's great remaining mysteries. How come some "living" Christmas Trees *don't* after they are planted.

A while back I spent some time with the grower who produces a large number of living Christmas trees in the Southwest. He supplies all the large, orange-colored, box hardware stores from Arizona to Texas. That is a lot of trees. His tree of choice is the Eldarica pine which is also known as the Afghan, Mondale, Goldwater, Arizona, Desert, and Quetta Pine. His estimate was that they lose less that 5% of all the pines they pot up.

I have also spent a lot of time talking with people that have bought and planted these pine trees. Although I haven't kept an exact total, I would guess that the mortality rate of the planted trees is much higher. Why is there such a difference in these two estimates?

Here are a couple of possibilities that I have considered.

Possibility 1. The grower is lying through his teeth about his success rate.
Response. That is not exactly in keeping with the holiday spirit. Besides, the man who started the operation is an honest man (U of A and Washington State Grad) who has extensive experience in agriculture.

Possibility 2. The trees are drying out in the house.
Response. This is a possibility, but most people keep their trees moist while they are inside by putting trays or blocks of ice on the root ball

every few days. Besides, these trees that have been in the cold are pretty dormant. They don't need as much water as a growing tree.

Possibility 3. Trees are damaged during planting.
Response. This is probably the biggest culprit in tree loss. Most Living Christmas Trees are field grown, then harvested and put in containers. Unlike nursery produce stock which has all of its root system in the bucket, field grown stock loses most of the root system when it is removed from the ground. Plants that are left in the bucket long enough to re-root seem to have a higher viability than those that are replanted right away. That would explain the difference in the success rates claimed by the grower and the people who buy them. These trees need to be treated differently than standard nursery stock when planting. Here is a quick guide.

Planting Living Christmas Trees

1) If your tree is wet, let it dry before planting. This will hold the dirt around the root ball together better.

2) Dig a hole 2-3 times wider than the diameter of the bucket and as deep as the root ball.

3) Put the tree in the hole - bucket and all. Always handle the tree by the bucket, not the trunk.

4) Cut the sides of the bucket away to remove them; and, it is OK to leave the bottom on. It doesn't affect the health of the tree.

5) DO NOT REMOVE BURLAP AND/OR WIRE FROM THE ROOT BALL. It is probably the only thing holding the dirt together at this point.

6) Backfill the hole using the dirt that came out. No amendments are needed.

7) Cover the top of your excavated area with 3-4 inches of mulch.

8) Water deeply.

9) Have a good holiday season.

The Big Money

(Jim's Note- Dear Readers, Puppy Trainers, and others who peruse this paper while completing their daily functions. Thank you for allowing me an undeserved break. I took off January to work on some projects which of course didn't get done. Then February came around and we had a high tech meltdown here at the old casa. Now, like the annual plants that are starting to appear after these nice rains, we're back.)

Psst? Want to make some big money in agriculture? No, not that. It is still illegal.

These days there just might be big money in native plants. When I first started working with native type plants some 25 years ago there wasn't much available. Those days Autumn Sage, a flavor or two of Texas Ranger, and Desert Willows, usually with washed out pink flowers, were all that were found in the native plant section of your local nursery. If you wanted something different for your landscape, you pretty much had to grow it out yourself.

Times are good these days. Native oaks? You can usually find three or four species if you know where to look. Every nursery has a mess of southwestern salvias in various sizes and colors. Native grasses are container grown for ornamental planting. Even gnarly shrubs like Cliff Rose and Apache Plume are produced on a large scale. I hardly have to do any growing on my own anymore because so much good material is available. However, there are still a few plants out there that have good potential as landscape material for those of us in the higher arid areas of SE Arizona. These aren't being touched by the large growers. Here is your chance to get in on the bottom floor and discover some plants with amazing economic potential.

I would start by growing one of our native penstemons that is found on the limey hills around here. It has a purple flower and thin grass-like leaves. It hopefully will cover some of the hills around here this spring. Don't waste your time looking for it in most of the "Wildflowers of the U.S." type books. Its range is too limited to be included (or we aren't considered a part of the

United States by the authors). The botanical name is Penstemon dasyphyllus. Sorry, but it doesn't seem to have a common name. I've collected seed from this plant and given it to some good growers in Tucson. Unfortunately, it struggles with the heat in containers there. That shouldn't be a problem locally. This plant would work in a variety of locations but would especially like our rolling grasslands.

If I could have just one plant to add to my arsenal this spring it would be Rock Jasmine (AKA Sonoran Trumpet or Sonoran Jasmine). For you botanical types it is Macrosiphonia brachysiphon. This plant is found in the wild at about 4500'-5500' in Santa Cruz, Cochise, and Eastern Pima counties. Rock Jasmine is a low growing ground cover with white trumpet shaped flowers. It is kind of scraggly in the wild but might look better under cultivation. Besides you don't grow it for the foliage. It is the fragrance of the flowers that is outstanding. Regular jasmine has the allure of a decomposing javelina by comparison. Let me illustrate its power.

Several years ago my friend Seth Everhart and I were out shooting. Now these days Seth is a very happily married man with a steady job and even a responsibility or two. Back then Seth was fancy free with three things and only three things on his mind. Hunting, guns, and Ford trucks. Anything else was superfluous.

Any rate, we were out in the National Forest burning some powder when we noticed a light sweet scent. Believe it or not, once we noticed the fragrance we lost all interest in shooting and searched until dark for the source. We never did find it as I think it was carried some ways down the canyon by the cooling night air. I know what it was, though, that made us forget what we were doing. It was Sonoran Jasmine calling to us like the Sirens.

The Really Big Money in Pinus

Last week we talked about a couple of native plants that could be cultivated for landscape purposes. The biggest gap to be filled in the native plant pallet would revolve around the genus pinus, the native pines. There are few, if any, that are readily available. Once again this is kind of a numbers game. The nursery industry is oriented to the giant population centers of Tucson and Phoenix. Most native pines aren't trees for the low desert and they aren't going to commit large portions of their growing space to a niche market like us. They make their big money in large scale production and "one size fits all" type plants. The best example of this is the Eldarica Pine.

I'd guess that over 90% of all pines planted from Phoenix to the Mexican border are Eldarica pines (Pinus eldarica). They grow fast, sometimes 2'-3' a year, and take our wind. Single digit temperatures don't affect them and neither does 100 degrees plus. This species is native to Afghanistan, Pakistan and a few other "stans". The growth habit is upright and they are usually about twice as tall as they are wide. They do well in our area, but in my personal opinion, lack the visual appeal of most native pines. In other words, they just are kind of boring. However, they work in most elevations and grow fast in containers. Therefore, they are produced on a huge scale. Now here are some examples of pines that would work as well, be more aesthetically interesting, and if you grew them on a limited basis, you would have the market cornered.

My first choice would be the Mexican Pinion Pine (recently changed to Pinus discolor). This native pine is a close relative to the Pinion Pine that produces the edible pine nuts. The nuts on the Mexican Pinion are edible but much harder to crack. The meat has a stronger pine flavor, too. This tree would make a great screen or patio tree. It is kind of interesting that they don't seem to have been affected by the drought as much as the native oaks or junipers. They have short needles in bunches of threes.

Chihuahuan Pine (Pinus leiophylla) is found naturally at slightly higher elevations than the Mexican Pinion. It is unique and interesting because it has a yellowish tint to its needles when it is young. It would contrast nicely if grouped with other greener species of native pine. I have seen these plants thrive in Patagonia. Collecting seed may be a challenge as they take three years to mature in the cone.

Another pine that I have seen grow rapidly in Patagonia is the Southwest White Pine (Pinus strobiformis). This pine is a higher elevation species that doesn't seem to mind being planted lower. It has soft blue green needles and long skinny pinecones. It is the southwestern version of the White Pine that grows in much of the rest of the country.

The plant that always gets the most attention in the nursery is the Apache Pine (Pinus engelmannii). It has long needles (l0"-12") in bundles of fives. I have seen them double their height in containers in a year. They do great in the ground especially if given a little more water than some of the other pines. Apache Pine is basically a Mexican species that can be found in some of our Southern Arizona mountain ranges. Its needles are used in the making of pine baskets.

Before you get growing on some of these amazing opportunities, there is something to be aware of. Seed for almost all of the pines named above has been kind of scarce due to the drought. It is out there but you have to look in the right places to find it. Of course, there are other good plants that could be grown that we didn't mention. I could write about them but it might take up too much of your time reading about all of them. Time you could be collecting seed, mixing dirt, and potting up plants. And making the big money.

Catalog Complex

I'm not a conspiracy theorist. This is not a theory, it is the bonified truth. Seed companies and the Post Office are in cahoots.

Every year when the days are wet and dreary the brightly colored seed and gardening catalogs show up. Sunny and mild, no catalogs. Henry Field's new one showed up during the last storm. Since the weather pattern has changed and we dried out, nothing. I know there are cases of catalogs in the back room at the Post Office just waiting for the next storm.

When these brightly colored catalogs arrive (often referred to as the Burpee-Gurpee-Herpes Complex), are they selling seeds and plants? Not Hardly! They contain a promise of hope and potential. Everything offered is the best and gives off the most bountiful harvest ever. It's a lot like the first day of kindergarten. At some point the bell rings. A little reality is a good thing at this point.

It helps to be able to translate some key catalog phrases as they apply to the Southwest. "Takes full sun" doesn't mean it takes our full sun. Give it a break from the late afternoon sun. "Drought tolerant" could mean it went three days between rains in North Carolina. "Tolerant of all soils" means nothing here. Technically, what we have on top of the ground here is classified by soil scientists as "soft adobe". I once grew some tomatoes that came from Russia. They were advertised as drought and heat tolerant, good for alkaline soil, and disease resistant. They did grow well and produced nicely. Unfortunately, they tasted like damp cardboard.

When choosing varieties, talk with some of your neighbors and see what has done well for them. It's best to plant several varieties of tomatoes and chiles, as what did well last year might not do it again this summer. Conditions will be different. Look for local catalogs like Native Seed/SEARCH. Their varieties have been grown in the Southwest for years. This year I'm going with more "heritage" varieties. These are nonhybrid varieties that often have been preserved and grown out for generations. You know what? This year is going to be my best garden ever.

Tree Tomato, the Rest of the Story

A couple of weeks ago I bought a Sunday paper, The Red Star, and then pitched out everything but the sports section, comics and the weekly magazine. It was while perusing the last of these sections that my attention was drawn to an ad for the "Amazing Tree Tomato" with its promise of bushel baskets full of mouth-watering fruits. The thought occurred to me that it would be a good research project to get a couple of these plants (for the amazing low price of 2 for $6.98 plus shipping and handling) and enjoy the fruits of my labor. Well, I might have been born at night but it wasn't last night, so I decided to do a little checking up first. You are going to find this hard to believe, but the "Amazing Tomato Tree" just might not be all that it is cracked up to be.

To start with, the "Amazing Tomato Tree" isn't really a tomato. It is a plant from South America (Cyphomanodra betacea) which is in the Solanaceae family. In all fairness, that family does contain potatoes, egg plant, jimsonweed, Deadly Nightshade, and tomatoes, so it is related to the tomato. I have to guess that the marketing guys thought "Amazing Tomato Tree" would go over better that "Amazing Deadly Nightshade Tree." Gotta give them credit for that.

This plant grows 6' -10' tall. It cannot take any frost so it would have to be kept inside in the winter. (That was strike one against it in my book.) It takes 18 months to 2 years for it to set fruit (strike two) then should continue to bear for 4 or 5 more years. The fruit look more like red egg plants than tomatoes but the real question is how "mouth watering" are they? If they really taste good who cares what they are technically classified as. Last week I made some Texas Caviar which had beans, not fish eggs in it. We all know oysters from the sea aren't related to oysters from the mountains. It is the taste, not the botany that counts. The text in the ad claims they are "mouth watering" but according to a statement from the Purdue Dept. Of Horticulture the "flavor suggests a mild or under ripe tomato with a fairly resinous aftertaste." Did they say "fairly resinous aftertaste?" I believe that would be strike three.

But wait; there is more to this amazing offer. Upon further research I found one source which was nothing but complaints about the outfit that sells the "Amazing Tomato Tree." In fact, it had over 20 complaints from people that sent in their money, had their checks cashed but never received what they had paid for. Of course after reading what the things supposedly taste like they might consider themselves lucky, but we still have to call that strike four.

If anyone has tried the "Amazing Tomato Tree" I would sure like to hear about it. I am glad to not be out the amazingly low amount of $6.98, mainly on principle. It seems that many of the best choices are those we don't make.

Special Bonus Feature: The world's best producing tomato plant is at the Epcot Center in Florida. It is a specially imported Chinese variety that is grown under very controlled conditions. So far it has produced over 20,000 golf ball sized fruits. Now if you could just get two of those for the amazingly low price of...

Sudden Oak Death

Back in July I had an Emory oak that just up and died after a couple of rains. Other oaks and junipers had headed south during the year but most of those had been in high risk areas like west or south facing rocky slopes. A few were next to the road so that their zone for water collection had been severely limited by compaction. This Emory oak was different. It was in a slight depression that collected water but had good drainage. During this drought it should have been a spot that promoted healthy growth. Even more puzzling was the fact that it died after the rains had started. I began to wonder if there was something else up. It was time to get some outside help.

A call was made to Linda Kennedy (from the Research Ranch), who I usually call when I have a hard plant question. Linda said that she wasn't aware of anything specific but the fact that the tree died after the rains started, and kept its leaves on, led her to believe that it may have been a fungal problem.

Right after I talked with Linda I heard from another friend who had just been reading about a severe disease that was wiping out a significant number of oaks in California. The symptoms were consistent with the Emory oak that had died here. The disease is simply known as Sudden Oak Death. It is getting a lot of attention across the country.

Sudden Oak Death is caused by a fungus-like organism called phytopthora. It is closely related to the fungus-like organism that caused the Irish Potato Famine. It was first noticed in Coastal California in 1995 after a wet year. It is very contagious and hard to treat. Although first confined to several counties in California it has now been found in Washington, Oregon, British Columbia, and Europe as well.

One of the possible ways that Sudden Oak Death may have been spread is through contaminated nursery plants. A large well known nursery in Southern California was found to have diseased plants. That nursery shall remain nameless. OK you twisted my arm. It was Monrovia. They had to destroy hundreds of thousands of dollars worth of plants.

The symptoms of Sudden Oak Death are pretty easy to see. A tree looks good one day and most branches are dead the next. It dies with its leaves on. There may be some brown ooze seeping from small holes near the base. The infected wood on the inside often turns a red color.

Shortly after I was made aware of Sudden Oak Death I went over to a neighbor's house. She had a beautiful old oak that had died. It was probably 35' tall and 40' wide. You couldn't wrap your arms around the trunk. This tree had been growing at the bottom of a canyon and was now dead as an anvil. It died with its leaves on. Several other smaller oaks around it were also dead. Could Sudden Oak Death have spread to Southern Arizona and is it starting to kill off our oak woodlands?

To be continued next week.
(Actually, that would be really rude to leave an important question hanging like that. The answer is no, and I will explain why next time)

Sudden Oak Death   (continued)

We left off last week talking about Sudden Oak Death, a serious threat to the oak populations in Northern California. Just to review, a tree on my property had died and was showing some of the symptoms common to that disease. They included dying with the leaves on, and a brown ooze weeping out of cracks in the base. This was a good time to call in a real expert.

Dr. Mary Olsen is the University of Arizona's Extension Plant Pathologist. If her name sounds familiar to you, she is the one who helped us figure out the Tomato Russet Mite problem that was ruining our crops a couple of years ago. She is the top gun in this part of the world for dealing with plant diseases. Dr. Olsen thought it was worth having the tree analyzed to see if it could have died of Sudden Oak Death. We certainly don't want it to get started in this part of the world.

We cut the tree down and noticed a hollow center. Dr. Olsen referred to that as Heart Rot. I am sure that I said something intelligent like "Isn't Heart Rot a punk rock love song?" Some of the wood on the interior was red in color. This is another symptom of Sudden Oak Death. We cut some samples away to be examined. About a week later we got word back. This tree did NOT die of Sudden Oak Death. Dr Olsen told me that it is very unlikely that we will be bothered by this disease here in Southeast Arizona. Why? Sudden Oak Death needs cold, wet conditions to flourish. With the current drought it would be almost impossible for the disease to take hold here. Mark it down; there is something positive about the current lack of moisture. OK then, what killed this tree and why are others dying? And how about the symptoms? Dr. Olsen told me that there are about 20 fungi or fungi-like organisms that attack our native oaks. Many produce similar symptoms to Sudden Oak Death. Three or four of these cause the Heart Rot that is common in many of the oaks that have died.

Ordinarily trees are strong enough to deal with the disease but the infection coupled with the existing stress caused by the drought, turns out to be a lethal combination.

You can see the same type of situation happening in the conifer forests of the higher elevations. Millions of pines have died. The Bark beetle is being blamed. But the Bark beetles are just the final nail. In years of "normal" rainfall the pines are able to produce enough sap to close the holes caused by the beetles. However, in drought conditions the trees can't cope. Stressing the trees allows other factors, which they could usually handle, to kill them off.

So now we know why the oaks are dying here in Southeast Arizona. Can anything be done about it? Yes but only on a limited scale. If you have oaks near your house you might give them an occasional deep watering. That should help. The bigger question is: is this drought going to have a major effect on the plant life we see around us today?

And the Winner is...

Continuing the long-standing tradition, begun last year, we are proud to announce this year's "Plant of the Year" award winners. This highly coveted honor is bestowed in five categories to the plants that performed the best in this year's specific growing conditions. Technically speaking, you would describe this year's growing conditions as having stank. Wind, drought, wildfires, and grasshoppers made 2002 the kind of year that sent the original homesteaders of our area on to California or back to where they came from. And, now the awards.

Tree of the Year: The heat and dryness didn't hardly faze the Native Mesquite, this year's winner. Even when the prickly pear cactus showed signs of stress, the mesquites looked great. They even produced a bumper crop of seedpods, which probably helped to keep some wildlife alive. Just to keep things clear, it's the Native Mesquite or Texas Honey that works here. Most of us are too cold for the South American Hybrids.

Shrub of the Year: Sorry no award given in this category this year as no single species stood out. While pondering this category I talked with local landscape expert William Kunk. William was also hard pressed to pick a single standout but said his vote went to any of them that withstood grasshopper pressure. He mentioned Tuscan Blue Upright Rosemary and Culinary Sage as a couple that performed well. Some of the other sages did well also but none towered above the rest. Better luck next year shrubs.

Groundcover of the Year: No doubt about, it the most impressive performance of a groundcover this year was by the native grasses after the wildfires. Despite a summer rainy season that rated somewhere between poor and mediocre, the native grasses came back to provide erosion control, feed and cover. I was walking in the Canelo Hills the other day and it was hard, in some spots, to tell what was burned from what wasn't because of the grass cover.

Annual of the Year: This was really close. Portulacas (Moss Rose) were as colorful as I've ever seen them. Their colors remind me of a Mexican fiesta. The hotter and dryer it was the better they looked. On the other hand Hummingbird Sage (Salvia coccinea) began its deep red bloom in May and in some places is still blooming in December. At my house it self seeded and came up between bricks and under rocks. The hummingbirds ate it up. On the basis of longevity, the award goes to the Salvia.

Garden Crop of the Year: By a landslide, it was the year of the cucumber. Those who planted them cuked early and cuked often. Most of the vendors that set up at the Growers Market had cukes to sell. At one point there was enough of a surplus to drive the price down to "as many as you want" for a dollar. Hopefully that will happen to tomatoes and watermelons next year.

Well, that's the awards for this year. You could do a lot worse than planning your landscape around some of these winners. They are some of the hardiest of the hardy. If they can make it through '02, they can probably make it through any year.

Plants of the Year   2003

Folks, thanks for being patient with me while I took some time off from writing this column. I am not saying that it was "well deserved time off" but I took it anyway. Now it is January and it is dry so things haven't changed too much from last month or even last year. But, being it is a New Year it is time for our annual "Plants of the Year Awards". These highly coveted awards are given out to the species in each category that have performed best over the last twelve months. Of course, past performance is no guarantee of future success but that is what keeps things interesting. If no species sticks out in a particular category then we will wait to next year to award that category.

This year a large portion of the judging criteria was based on personal experience rather than discussions with the knowledgeable plant people I have the good fortune to know. This happened for two reasons. First, my wife took over watering around here and she has a heavier hand with the water than I do. I got to see how some of our plants can perform, and not just survive, with adequate moisture. Secondly, I haven't got to town too much lately so I haven't been able to do much polling. Here are the awards.

Annual of the Year goes to our own native sunflower, Helianthus annuus. These plants line our roadsides during the late summer rains. Each plant has the potential to get about 10' – 12' tall and can be covered with hundreds of deep yellow blossoms. Butterflies like the flowers and birds, especially doves, eat the seeds. Of course when the State Dept. of Transportation poisons them, they usually don't do so well. You might want to remember to send your "greetings" to the DOT early this year and tell them to keep their sprays off of our native annuals.

We have a runner up for Annual of the Year this year and it is Salvia coccinea. This plant is also known as Scarlet Sage, Texas Hummingbird Sage, Lady in Red, and my favorite, obliviously dreamed up by some kind of marketing genius, Forest Fire. Although it has placed in the Annual of the Year category, this Salvia can be perennial in slightly warmer areas. This year it came back from seed all over the place at my house, even in the cracks of bricks. It grows about 2'-3' tall and has lots of scarlet (red) flowers on spikes that attract hummingbirds. Most of the time it reseeds itself and not only gives several generations a year but will come back from seed the next year.

Perennial of the Year goes to Echinacea purpurea, also know as Purple Coneflower. This is the same Echinacea that is used for medicinal purposes.

This year, in its second growing season in my backyard, Echinacea bloomed constantly from mid-April through September. I don't believe there was a single day that it didn't have at least several flowers on it. The flowers are about 2 ½ inches in diameter with pinkish purple petals on them. Much of the time the flowers were hard to see though because the plant was covered in butterflies. Monarchs and Queens seemed to favor them over other plants that were in bloom. Purple Coneflower is a Plains native that likes a little more water than some

of our native perennials. I should mention that the plants I grew were planted in completely native soil, mostly clay and rock.
To be continued next week.

Plants of the Year 2003 (Continued)

We left off last week by naming Purple Coneflower as the Perennial of the Year. Little did I know that it had been named Perennial of the Year for the whole country in 1998. I really don't like being in agreement with other parts of the country but 6 years have passed so I guess that it is OK. I hope not too many of you had problems sleeping while you were waiting for these announcements to continue.

The award this year for the Shrub of the Year goes to ….. nothing. No one particular shrub stood out as having done spectacularly this year. I could have presented it to Autumn Sage (Salvia Gregii) as it always does well. In fact it is probably our single most dependable plant year in and out. However we are looking for outstanding performance this year, not lifetime achievement. Sometime when there is really nothing going on we can have a Lifetime Achievement column.

Vine of the Year goes to Coral Honeysuckle (Lonicera sempervirens). This plant used to be called Evergreen Honeysuckle. Coral Honeysuckle was in some state of bloom from the late freezes of March until the early freezes of October. It has tubular coral flowers, about 2" long, in clusters. (This seems to have been a good year overall for tube- shaped reddish flowers.) This plant does best with some break from the late afternoon summer sun, but heck, don't we all? Coral Honeysuckle will keep most of its leaves until the temperatures get down in the low teens. As you have probably figured out, it is a good plant for the hummingbirds. This plant does so well here that it is often mistaken for a native. There is a species called Arizona Honeysuckle (Lonicera arizonica) that I have seen growing in the White Mountains that looks very similar.

Garden Crop of the Year, for the second straight year, goes to cucumbers. Partly, this was from a lack of competition. It was a pretty rotten year for both chiles and maters. Early in the year people actually complained about getting no zucchinis. Later in the summer the big hail storm hit and put a lot of folks out of the gardening business altogether. Despite all the problems, cucumbers were plentiful. And they were good.

Now let's wrap it up with the biggie, the Tree of the Year.

The winner this year is certainly a dark horse candidate. Most other years' late frosts keep this tree from even being considered. In many locations, if this tree comes through one out of four years you are lucky. A lot of people got lucky this year. Once you have had good ones right off the tree, buying them in large supermarkets just doesn't cut it anymore. (Red Mountain, in Patagonia, usually gets good ones.) We hope your crop was good enough to share with some friends. This year's Tree of the Year, the apricot.

Plants of the Year 2004

It seems that anyone remotely connected to the media puts out a "best of" list this time of year. It is supposed to be a review which shows great understanding into the events of the last 366 days. The truth is that most of these "best of" lists are put together by someone who has decided that it is more important to eat cookies and partake of holiday beverages than to create something new. So, here is my Best Performing Plants of 2004 list.

Groundcover - We have a native plant in SE Arizona that loves the heat and thrives with little moisture. It can be found by the side of the road blooming, starting in the early summer with 2" yellow flowers. With some supplemental irrigation it gets about 18" tall and 3' - 4' wide. This plant is good at attracting butterflies. A smaller and more compact version of this plant with brighter yellow flowers comes from Texas. Both varieties performed extremely well here last year. The 2004 groundcover of the year is Calylophus, aka Sundrops.

Vine - Several types of vines did well this past year. Tombstone Rose sure had a nice bloom last spring. Coral Honeysuckle (last year's winner) bloomed for most of the season and really drew the hummingbirds. Even some of the climbing roses had good flowering periods due to the long cool spring. However, our judging panel has very high standards. All of the above performed well but, that is their job. None were exceptional so no award is given in this category this year.

Shrubs - Again no award is given because no type of shrub really outperformed the rest. It was a decent year overall. The good news is that there was a tremendous increase in the amount of new varieties of "native" shrubs that were available to plant this year. Many of these were Southwestern or Mexican species of salvias. I wouldn't be surprised to see one of them come up a winner next year.

Crop - This was the easiest category to pick a champion in. 2004 was definitely the year of the fruit tree. Apricots, plums, peaches, Asian Pears, pears, and apples all came through with abundant harvests. Hope you enjoyed it because it might be a while before we see another year like this one.

Tree - This is the most prestigious of the awards and it was hotly contested as usual. Mexican Elderberry got off to an early and strong lead. The elderberry trees around Patagonia were covered in dark berries. If you were making wild elderberry jam, this should have been your year. However if you were like me and got there a week too late, the trees were naked and the birds were fat. The Elderberries were looking strong until...

Around midsummer a strange phenomena began to occur. Peaches started to appear where there had never been peaches before. Trees that had been barren for years, and trees that had never produced, all bore fruit. Peaches were given away by the sack-full. One time I parked my truck in front of the feed store with a sack of peaches on the front seat. My windows weren't rolled up and the doors were unlocked. When I returned someone had left two more sacks on the seat. Of course all of the peaches weren't great. Many were picked early and green to avoid breaking branches. We just weren't ready for peach success. The tart peaches made good jam. All of the white varieties were excellent tasting. 2004 will be remembered as the year of the peach tree. It was the year peaches were zucchinis.

Plants of the Year 2005

Once again, thank you for allowing me to take some time off. I like taking a break this time of year because basically, it feels good, and there isn't a whole lot to write about anyway. I might still be on break if it wasn't for a chance conversation that occurred one day, while I was pulling nails from a pile of used lumber in front of the feed store, between myself and Donna, the mild-mannered manager and sweetheart of The Bulletin. It went something like this...

Donna: "Hey, is that you? I didn't recognize you working. I've never seen that before. When are you going to get us another article?"

Jim: "Well, I've been kind of enjoying the break and besides, I haven't really had much to say. And, unlike much of talk radio, I believe that there is an obligation to say nothing when you have nothing to say."

Donna: "That's nice. I'll expect one for next week then."
Jim: "OK. No problem."

So folks, we are kicking off 2006 with the annual best plants review.

Overall, 2005 was deceiving when you look back at the numbers. Most of us ended up with something close to average historic annual rainfall. Though how it got here was anything but average. We had the hottest July on record and the latest recorded start of the summer rains. That was followed by, for most of us, an excellent August where we got a whole summer's worth of rain in one month. Since then we have been pretty dry. All that made for a tough year on plants and gardens. Let's give out some awards.

Annual of the Year – Although it looks like something that could be in an English cottage garden, this year's winner is tougher that it looks. This plant is small, around 8" x 8" and is covered with lavender, pink or white blossoms. It works in pots or beds and can take sun, partial shade and is fairly drought tolerant. Best of all, butterflies really are attracted to it. The Annual of the Year 2005: Pentas.

Garden Crop of the Year – For awhile it looked like this category would be a no show. With the heat and lack of rain it was all you could do to keep a garden alive much less make it productive. Then we got some moisture and things took off. What really took off for me this year were the chiles. (Since I am writing this, I can give awards based solely on personal experience.) In one picking, from just 12 plants, we filled a whole basket with green chiles. These weren't just any green chiles though: they were Santa Cruz Blues, a variety developed locally by Paul Thornburg. 2005 was the Year of the Green Chile. Did I mention I won the blue ribbon at the County Fair in the green chile category?

Groundcover of the Year – This was probably the easiest award to hand out. And it shows, like a lot of other things in life, how important timing is. This whole area was looking mighty brown in late July when the rains started. Usually it takes about a week of moisture before things start to green up. This year was different. A week passed and it was still brown. Another week went by and it didn't look much different. Was this going to be the year it stayed brown? Fortunately for us, heck no. It took a while for the grasses to get green but then they took off with a vengeance. I don't know if it was the near-death experience, but when the grasses went to seed, it was the best seed crop in recent memory. Most native grass seed is viable for almost a decade. This was the year that hopefully put enough seed in the (soil) bank to cover us for awhile. That is why the native grasses are the 2005 Groundcover of the Year.

(We will let the tension mount until next week when the Plants of the Year list will continue with the Shrub of the Year.)

Plants of the Year 2005 (con't)

I hope you all have been able to sleep while waiting for the announcement of the Shrub of the Year and the rest of the awards. It would be cruel to keep you in suspense any longer.

But before we reveal the Shrub of the Year, for the first time ever we are giving a runner up in this category. The runner up is a member of the Salvia family. More specifically it is a type of Little Leaf Sage (Salvia microphyla) from the Sierra Madre area of Mexico. What makes it unique is the flower size, which is about one and a half times bigger than most of the other Salvias that we use, and its long flowering period. It doesn't seem to get covered with blossoms at any one time, but always seems to have something going on. So this year's runner up is Salvia var. Scarlet Spires.

And the winner of the 2005 Shrub of the Year is Agastache var. Summer Breeze. This was the second year in the ground for this plant in my backyard and it really took off. It probably grew to about 3' x 3' and was covered for most of the summer with hot pink and orange tubular flowers. Yes, it had both at the same time. Hummingbirds and butterflies flocked to it. Summer Breeze managed to over grow and outshine the Echinacea that it was planted next to. I know we covered it before in the article "Penestemons are Over-rated" but most of the Agastache family does well here and draws a variety of wildlife, including native finches.

The winner of the Vine of the Year is no surprise. In fact it has won the award in several different years. It is a perennial winner. Coral Honeysuckle bloomed nonstop from March through September. It thrives in anything less than full sun and has coral tube-shaped flowers that, you guessed it, attract hummingbirds. Although it lacks the fragrance of its better known relative, Hall's Honeysuckle, the longer blooming period makes it a more interesting choice to plant. This plant belongs in our Plants of the Year Hall of Fame.

And now the moment we have all been waiting patiently for, the biggie, the Tree of the Year. This year's Tree of the Year is a native that grows mostly in the area around washes and riparian areas. At times its leaves have a distinct sweet smell to them. What earned it the highest honor this year was its blossoms. It bloomed over a longer than usual period possibly because of the strange weather we had in the early summer. The orchid-like flowers are light purple to burgundy colored and are sought out by hummingbirds and especially Pipevine butterflies. It is with great pleasure that we award the Tree of the Year honors to the Desert Willow.

Looking back on our list of selections, it seems that this year was a little different. Usually we choose plants that happen to have an unusually successful year like the time when apricots, which usually get ruined by late frost or wind blowing off the flowers, set a good crop. This year's picks are all good solid choices that perform every year. Most were chosen because they had an extended blooming period and attracted wildlife. (The kind of wildlife that is fun to watch: not the kind that eats your plants or pulls them out of the ground.) They would all be good choices to base the plants in your landscape around.

Well, thanks for checking out the Plants of the Year 2005 edition.

How to Plant a Rock

One of the best materials to use in the home landscape is a boulder. They look natural, offer protection to plants near them, and are very hard to kill. All plants, even cactus, need water, rocks don't.

The most common mistake that occurs when using boulders in a landscape is improper planting. Usually someone just backs the old pick- up in and rolls out the rock. Where and how it lands is good enough. Make the mistake of asking the proud owner what kind of rock it is and the response is usually "leaverite". As in "I'm gonna leaverite there".

Boulders should look like they are growing out of the ground. Before you place the stone dig out a place for it to rest. Place about one third of it underground. No edges should be showing on the bottom. This will help it appear as if the rock had always been in place. Put the rougher more jagged side where it will be seen better. It is more interesting to look at. The area at the base makes a good planting zone. Rarely do you see a rock standing by itself, with nothing growing around it, in the natural world.

Other parts of the world hold rocks and their placement in much greater esteem than we do. Japan in particular is famous for their appreciation of boulders. A particularly well shaped rock will be given a name (probably not "leaverite") and treasured like a work of art. I've heard of boulders being sold for tens of thousands of dollars in Japan. Wait a second. I'll bet most of you who are reading this are looking out your windows and thinking that there's gold in them thar hills.

The Conference Review

I just got back from the 13th High Desert Gardening and Landscaping Conference in Sierra Vista. This conference is the best single source for relevant information about landscaping, plants, and water usage in the higher elevations of our region. Of course there were many serious sessions about the use of native plants to attract wild life, water conservation, plant diseases, and the contentious water rights issues we will be facing in the very near future. However, I am going to avoid most of these and concentrate on a series of facts that I learned that made going to this event fun.

John White is the County Program Director for the Extension Service in Dona Ana County in Las Cruces, N.M. His talk was on chiles. John told us that there is more vitamin C in one chile than there is in six oranges. Leaves from the chile plant are edible and can be used in salads. A large amount of the chile that is raised in New Mexico is used for dyes and not eating. Red and green chiles have the same amount of heat but the red has more sugar which helps to tone down some of the burn. The next time your mouth is burning, reach for some sugar rather than water. And most important of all, red chiles are fed to flamingos in zoos to give them their pink color because they can't afford to feed them the shrimp that would create the coloration in the wild.

Matt Johnson, author of "Cacti, Succulents, and Unusual Xerophytes of Southern Arizona" told us that the Claret Cup Hedgehog is the only U.S. cactus pollinated by hummingbirds. Our own world renowned agave, the Huachuca Agave can live up to forty years. Another local agave, the Small-flowered Agave, is one of the few that lives, up to four years, after it flowers.

Apologies to Barbara Skye Siegel, instructor at the Sonoran Arthropod Studies Institute. She warned us that her talk about moonlight gardens would be "kind of arty." As a person whose main influences in landscape design have been Fred Sanford and Larry the Cable Guy, I decided this would be a good time to go unload my truck.

Kazz Workizer helpfully shared some of the knowledge that she gleaned from Cheri Melton's, owner of Planthoe Garden Factory, talk. It seems that when butterflies gather around pools of water their real intent is not the liquid but the surface salt. It helps the males raise their sperm levels. When the females emerge from their cocoon, and are waiting for their wings to harden (not dry off) they produce an irresistible scent to the males. (Remember, they have been off with the guys, downing salt shots.) The male butterflies are drawn to the females who can't fly off yet and a whole lot of butterfly breeding takes place.

Dr. Randy Norton is a University of Arizona Soil Specialist. He told us that one gram of clay can have a surface area of 10 to 1,000 square meters. Arid land soils usually have only 1% to 2% organic matter. If you are having problems with salt build-up, run your irrigation 2 to 3 times longer than usual once a month to leach it out.

Cochise County Horticulture Agent Rob Call spoke on weeds. He told us that a single Pigweed (aka Careless Weed) plant can produce 400,000 seeds which can remain viable for up to 20 years. Another weed, Cocklebur, was the inspiration for Velcro. A Swedish scientist was struck with the idea for Velcro while removing them from his dog's fur. Another weed expert Dr. Larry Howery told us about Star Thistle. This invasive thorny pest was unknown in California until 1958. It now covers 20 million acres, making them unsuitable for recreation or livestock.

Finally, there was a session with a panel of experts to answer questions and debunk myths. Among the nuggets revealed in that hour was that there is no scientific evidence that Marigolds will repel bugs from your garden. They may, however, attract good insects. Also, gopher vibrators and other sonic rodent eradicators are useless. Again no evidence they do anything. The biggest myth to fall was one concerning Vitamin B_1 and Superthrive. When both are analyzed they turn out to be basically nothing more than watered down fertilizer. And I know this might be hard to believe, but the Internet rumor about mulch from New Orleans and Formosan termites is totally false. Commercial mulch is not being made from debris in New Orleans.

So don't spend your money on gopher vibrators or root stimulants. Save it for the next High Desert Gardening and Landscaping Conference.

The Return of Five Easy Questions

(For those of you who have never seen it, this sophisticated technique lets us do two things. We can cover a wide variety of topics and we get a column in by the deadline when we don't have a lot to say about anything in particular.)

1) Have you gotten many suggestions to add to the list of water saving tips that you wrote about?

So far we have received a few. Rob Horsmann recommended putting a 5 gallon bucket in the shower to collect the water that runs while you are waiting for the hot water to arrive. You can also save water in the shower by running the water to get wet, turn it off to soap up, then turn it on again to rinse off. This should save over half the usage of a regular shower. (Some people, usually those that have served in the Air Force, Army, or Marines, refer to this type of sanitation as a "Navy" shower. We are not going to do that here.)

Another suggestion for saving water was to have fewer kids. Although the potential water savings is huge, in my case, it is a little like closing the barn door after the horses have already left.

2) It seems that most of the columns these days are the "drought this" or "water conservation" that. The overall tone is kind of grumpy. What happened to the ones that actually had useful tips about planting or gardening in our unique part of the Southwest?

Ok, ok you asked for it. Go plant some cilantro and dill.

3) Have you seen anything interesting lately?

I am really glad you asked this question. Last week I split open a pumpkin that was left over from last fall to feed the chickens. Inside the pumpkin, some of the seeds had sprouted and begun to grow leaves and roots. We have seen that before so it wasn't too surprising. What was different was that the leaves were dark green like they had started to photosynthesize and not pale from lack of light. I haven't figured that one out yet.

4) Have you seen anything interesting lately that doesn't involve chickens or pumpkins?

Obviously you must be referring to the sight of Manzanita flowering now. In a "normal" year (read, one not racked by historic drought) they flower sometime around February, not in April or May. It takes lots of energy to blossom and make fruit, and last winter there wasn't enough moisture to fuel the flowering. In fact there wasn't enough water to sustain the plants and many, especially on west or south slopes, have died. In March we got a bit of moisture, some of it in the form of snow, and a plant has to do what a plant has to do. It will be interesting to see what happens now. I believe that Manzanita is insect-pollinated. In February about the only pollinators out are bees. Now there are lots of potential pollinators. Will this make a difference? Will the berries even form in hot conditions? It will be fun to watch this one.

5) How come this is called "Five Easy Questions" when there are only four questions?

Well I guess that wraps this up. Thanks for taking the time to read this. We'll be back again next week.

COMPANY

WHY I'M LIKING THIS WEATHER   p. 110

THAR SHE BLOWS   p. 111

CALICHE: LOVE IT OR LEAVE IT   p. 113

PLANTING, THE SEQUEL   p. 114

WHY GO NATIVE   p. 115

DO SOMETHING ABOUT IT   p. 117

O SAGE, CAN YOU SEE?   p. 119

FIELD GUIDE TO LOCAL ROADSIDE WILDFLOWERS   p. 120

A QUICK COMPARISON   p. 123

A DEEP PHILOSOPHICAL DISCUSSION   p. 125

NON-NATIVES THAT MAKE THE CUT   p. 126

USE IT TWICE   p. 128

JIM'S (WATER) WORLD   p. 129

THE CONTINUATION OF JIM'S (WATER) WORLD   p. 131

L.E.M.ING   p.133

JUST NUKE 'EM   p. 135

A FIELD GUIDE TO WHAT IS EATING YOUR PLANTS   p. 137

Why I'm Liking This Weather

(This was written back in late spring of 2002. The drought was in the early stages and still kind of a novelty)

We're on a record setting pace. Today in Tucson, they will have gone 84 days without a single drop of rain. This is the second longest completely dry spell in their recorded history. A week from now they probably will have gone 91 days, which breaks the all-time record that goes back to 1909. The measly couple of drops we got about a month ago keeps us in SE Arizona out of the record books, but we've basically gotten no measurable moisture for over four months. According to my friend and weather watcher Paul Thornburg, this puts us on a pace with 1972 when there was also no moisture for that long. I was visiting someone in Sierra Vista who had been given a digital rain gauge for Christmas. He had recorded one one-hundredth of an inch this year. I don't believe you need a digital rain gauge to tell you that it is dry out there. Now, I'm not going to say that every cloud has its silver lining because 1) there aren't any clouds to speak of, and 2) the "linings" are more like aluminum foil than silver. Still, if you look hard enough there are some good things about these conditions.

Gone on a picnic lately? You can probably leave the bug repellent home. There are very few mosquitoes out there. Usually we spend as much time swatting mosquitoes as eating. So far the dry conditions have limited their populations. For most of us the fly and grasshopper numbers are down, too. In the nursery, we've had almost no aphids or floral thrips this year. Guess bugs like moisture, too. I have to wonder though if massive fires in the ancient days didn't also keep grasshopper numbers down.

Might have the time to go on your picnic because you're not spending hours pulling weeds. Most other years we're fighting tumbleweed right about now. Remember all the mustard last year? Basically there are no weeds this spring. Unfortunately, it's a temporary time-out. As soon as there is moisture the seeds will sprout and we will be back hoeing again. Enjoy it while you can.

Not everything looks bad right now. Mesquites, Canyon Hackberry, and Desert Willows look good. Some of the riparian trees such as AZ Ash and AZ Walnut aren't showing signs of stress either. They are just a little thinner than usual. This is giving us an excellent lesson as to what holds up during extreme conditions. I doubt Purple Leaf Plums or Globe Willows will survive without a lot of help. If you have established ornamental plants a couple of good soakings should keep them alive until help arrives.

There is a beneficial exercise component to these dry conditions also. Most of us are getting a good workout for our neck muscles by looking up to the south for clouds about 20 thousand times a day. We'll all have neck muscles like Brahma bulls.

In the future our kids will have bragging rights. They should start practicing for when they are adults right now. "Drought? You call this a drought? This ain't nothing. Why I remember when I was a kid back in '02 we had a dry spell that…"

OK, I know I'm stretching it a bit trying to find good things about these conditions. If you are having to buy feed for your livestock, or are constantly watering to keep things alive, you might disagree with some of my conclusions. One thing we will agree on, though. The best thing about the drought is how good it's going to feel when it's over.

Thar She Blows and Blows and…

I don't know why but I've had wind on my mind, and every other part of me, lately. After last year I swore I was giving up thinking about weather but somehow I just can't help it. We all know it's going to be windy in the spring but the question that made me fall off the weather wagon is: why?

Not being a meteorologist I called my friend and local property owner Chris Reith, who is a meteorologist with Air Quality Control. Chris explained that there were several factors that played into the equation for wind, but the most important was warm and cold air masses coming together. Differences in air temperature create air movement and the greater the difference, the stronger the winds will be. In the spring our days get longer and warmer. The rest of the country can still be pretty cool. Today is a good example. The highs in Phoenix are supposed to be around 90. Salt Lake City will only get to about 30 due to a cold front out of the northwest. The meeting of these two very different air masses is creating a lot of wind. If large areas of the country have similar temperatures there probably won't be much wind. When we are warm but Colorado or Utah is getting snow, you better keep small people and pets inside.

Even if there weren't a conflict of air masses we would still enjoy breezes most days in the spring due to convection currents. The longer and warmer days play a big part. As the air on the ground warms up, it begins to rise. The cooler air flows in to replace it. This cycle is your basic convection current. In our case, the air that comes in to replace it comes from the southwest, which is the prevailing direction of our winds.

A final reason that we are blessed with all this potential wind energy (trying to put a positive spin on it) has to do with our elevation. The same situation that makes us cooler in the summer kicks our butts in spring. We have a lot more wind in Southeast Arizona than they do in Tucson or Phoenix. My friend Peter Gierlach (of "Petey Mesquitey" fame) likes to say that this is the only part of the country where vise grip pliers double as clothes pins. At ground level there are a lot of obstructions that slow down or slightly change wind direction. Mountains, hills, trees, buildings, large animals, etc. all create friction that slows down the wind. Higher up there is less to create friction so the wind gets a running start. The same wind measured at 15 mph at sea level is probably traveling at about 40 mph at 4000 feet. In oversimplified terms, the higher you are the faster the wind blows.

Well, there you have 3 factors that, when mixed and matched, combine to give us our spring winds. Recognizing the causes doesn't slow down the wind, but it is the basis for understanding. It is taught that understanding is the first step in living in harmony with things you may once have despised. Personally, I think that is a bunch of hooey as the wind is blowing hard enough today to form white caps in the toilet. How long do the winds last? They will be around until the summer really begins to heat up. Then, I'll guarantee you, someone will say "sure could use a breeze today."

Caliche: Love It Or Leave It

Recently, I had the good fortune to attend a geology workshop at the Appleton-Whittell Research Ranch. The course was taught by local geologists Jan Rasmussen and Sandy Kunzer. It was as interesting a two days as I've had in a long time. You really have to appreciate how geologists think. Something that happened 60 million years ago is a "recent event". Makes you wonder what they would mean if they said they would be a "little late" for a meeting. By plant standards, a 200-300 year old tree is "ancient". One of the subjects covered briefly was an understanding of how caliche is formed. Caliche is that beloved layer of whitish dirt-rock that occurs near or on the surface of much of the arid Southwest. Sometimes people refer to clay as caliche or claim that it is acidic. It's neither. Caliche is chalk-like and very alkaline.

Caliche starts to form when rain, which is acid by nature, falls on limestone. The acid rain dissolves some of the calcium in the limestone and runs off into the ground. This deposit builds up over time and becomes caliche. A geologist might refer to it as calcium carbonate. The question is what to do with it if you want to plant in that area.

The problem is too big to try to change the soil by amending it. Some local planting guides will recommend digging through it. Sounds good on paper. Unfortunately, some caliche layers can be very deep. I've looked down leach field trenches (10'-12') and not seen the end of it. I don't believe you are going to dig through that.

Probably the best idea if you need to plant in that area is to use alkaline tolerant plants. Cliffrose (Cowania stans) is a shrub that thrives in white dirt. That is where it prefers to grow. Emory Oak, Evergreen Sumac, One Seed Juniper, and Beargrass all occur naturally in caliche soils.

Sometime the rules are broken. Years ago I planted an apple tree near the crossroads in Sonoita. Almost pure white dirt. It should have struggled and shown deficiencies. Instead it has thrived and bears almost every year. From a nurseryman's perspective it doesn't make sense. Maybe I should go ask a geologist.

Planting, The Sequel

It has occurred to me that some people might want to supplement their yard full of rocks with plants. So here's a guide to proper planting technique. Everything except flower gardens and vegetables can be planted this way.

Start by measuring how deep your root ball is. That is how deep your hole should be, no deeper. The width of the hole is 3-5 times the diameter of the root system. The worse the digging the wider the hole needs to be. Remove the plant from the container. If the root mass seems to be matted slice it down the sides with a pocketknife. When you place the plant in the hole make sure the top of the root ball is even or slightly above the existing grade. Whatever dirt comes out of the hole goes back in. Adding organic material as a part of the backfill mix is not necessary. After the plant is in the ground, cover the entire planting pit with 3-4 inches of mulch. You can use straw, hay, bark, or whatever will keep the moisture in and won't blow away. Finally, give it a thorough watering. In a day or so water again. This helps not only to water the plant in but also to eliminate air pockets in the soil.

What about the "California" method of planting that we all used to use? This is where all plants got a 6'x6' planting pit and you tried to improve the soil by adding mulch, peat moss, soil sulfur, and dead fish. Research by the University of Arizona and others has shown that adding organic material doesn't help and in the long run might even be detrimental. To be successful, the root system must grow out beyond the planting hole. If too much organic material is added the roots may just keep circling the hole and never leave it. Kind of like having a planter in the ground.

If you have a planting method that works for you and that you are comfortable with, stay with it. Here's something to think about though. I planted two apple trees side by side. One got the "California" treatment and the other was planted using the method with no amendments. I watched them for several years and there was no difference in growth. In other words I got the same results for less work and cheaper costs. What do you think is better?

Why Go Native

I never knew that the old term "January thaw" meant all 31 days, but apparently this year it does. The warm weather makes everyone want to get out and get started on their spring chores. I even fired up the rototiller and did some work on the garden. (Actually it was work that supposed to be done last fall, not this spring. Wouldn't want anyone to accuse me of getting ahead of myself.) This winter warm spell gives you the feeling that all gardens will be productive, all fruit trees will bear, and your shade trees will grow like weeds. It's a time to philosophize, not sweat. Here's a short quiz to get your mind wandering. Of course the answer is at the end of this article. Be careful, it might be tricky.

Q. What is the best reason to go native with your plants?
a) You get to run around your property wearing nothing but a loincloth and a smile.
b) It means never having to weed or prune.
c) You get the most results for your effort and expense because you have chosen plants that thrive in your conditions with the minimal amount of help.
d) All of the above.

Before we go native it would be kind of nice to figure out what plants are "native". Does it just refer to plants that grow here now? How about 100 or even 500 years ago? The definition preferred by botany geeks says that native plants were those that existed in an area before the arrival of that dastardly dude Columbus. That definition doesn't help us much in terms of landscaping so for our purposes, let's call native those plants that occur naturally in a given region and that thrive under the climatic conditions with little or no help. For us that includes many Chihuahuan Desert plants (salvias, hesperaloes, etc.) that work well here but are found a little to the east. What we really mean is native, or, Southwest adapted.

What makes these plants work here is their tolerance to our unique conditions including soil, wind, heat, cold, drought, and various combinations thereof. Our soil is known for it's high alkalinity-low fertility. It has very little organic material and very little nitrogen content. Other than riparian areas or cienagas there isn't much here that is green and leafy. Those kinds of plants like more acidic soil and more moisture, especially humidity.

Can you change the immediate conditions for your plants to make them more favorable for non-natives? It's probably more work than it is worth. Some older planting instructions will suggest using soil sulfur to lower the pH and acidify the soil. Research shows that this is probably a bad idea. Depending on your soil, it can take up to a pound of sulfur to neutralize one pound of soil. A good tree hole might have 200 lbs. of dirt come out of it. Eventually, we hope, your tree's roots will outgrow the hole to reach the native soil anyway. Better to pick something that likes our dirt to begin with.

The same principle holds up for our other "unique" conditions. If a plant can't hold up in the wind, don't waste your money on it. Pick up an Emory Oak leaf. It's not like northern or eastern oaks. It is leathery to the touch and holds up well to the wind. It has evolved to thrive in our conditions.

Native and southwest-adapted plants give the most return for the least effort. There is nothing wrong with the "outsiders" like fruit trees or roses but they do take more work to keep healthy. That is fine if that is your hobby. But remember, in periods of neglect, they will suffer more than the natives. My main landscape philosophy is that a landscape is to be enjoyed, not endured.

Now the answer to the quiz. All the answers are valid but we asked for the best. According to our philosophy, and it is that time of year, the best response is A. Go have some fun.

Do Something About It

I really was going to stop carrying on about the drought and what it is doing to the countryside. I really was, I promise. You could have been reading an article about roses, new varieties of daffodils, or exciting improvements in clay pots. But two things happened to set me off again. Sorry.

The first was an article by an Urban Extension Horticulturist about the old street trees dying in Tucson neighborhoods from lack of moisture.

The second was a conversation I had with a supposedly intelligent visitor from the East about the drought and its effects on the plants, fire conditions and water table here. Toward the end of our talk he paused for a second then asked me "so what should the federal government (Congress) be doing about the drought?"

At first I thought he had to be kidding or was jerking my chain to get a reaction. What was he looking for, passage of a law that declared the drought was over, cloud seeding, or just to form a committee to study it?

Now I know that the federal government funds agencies that are critical to fire containment such as the Forest Service and the Border Patrol, and I am very grateful to the folks that work for them, but the basic well-being of my property starts with me and not the government. I need to do what I can to keep my vegetation alive and reduce the fire risk, and I need to do it now.

As for all the boo-hoo-hooing about the street trees dying in Tucson from lack of moisture, here is an idea. Stop whining and water them! The theme here is it is time to do something about it.

The two areas that you might be about to do something about are, of course, fire reduction potential and plant survival. What is making this year different for fire than all the others? Just record drought, lots of dried out plant material from last year's decent precipitation, tons of illegal immigrants that start fires in the hills for warming, cooking or possibly diversion, and very little moisture in the vegetation that is still alive. If the long range predictions are correct, we are in for a warmer and drier spring while the Rockies to the north are expected to be cooler than normal. To me that means a windy Spring which only worsens the fire potential. Not too many years ago it seemed that a big fire burned several canyons in a mountain range or a couple of structures. Now fires take whole mountain ranges or even, as we have seen recently, whole communities. You have to wonder if a mega fire might someday go from range to range.

The first thing you should do for fire prevention is to educate yourself on creating defensible space around your structures. According to a friend that had a career in fire fighting with the Forest Service, the usual recommendation is to cut 50' of grass around your house. This year he said he would triple that. Recently, I bought a chipper and am in the process of grinding up all the dead branches that I can reach from the ground on the trees surrounding my house. Hopefully this will keep fire from climbing up the trees like a ladder and spreading. I have started close to the house and will work my way out with a special effort on the trees to the southwest, which is where the prevailing winds come from.

These measures will not stop a fire but will hopefully make it less intense and give us more time to have a chance to fight it. Now go pick up the phone and call the Forest Service or a local fire education agency and find out what you should be doing.

We have covered saving mature vegetation before. It is pretty straight forward. If you want to keep something alive, water it deeply, just outside the drip line (edge of the branches). Don't water at the base of the tree as most of the important feeder roots are not there. I am guessing that watering about every three weeks should get the job done. Remember, you aren't trying to grow the tree, just keep it alive until the natural moisture takes over, hopefully this summer.

O Sage, Can You See

Last week was a big one at the nursery for "sages". I'll bet 20 times I was asked "if I had sages" or heard "I didn't know that was a sage". The problem is that calling something a "sage" doesn't really mean too much. A sage is going to be something different depending on what part of the country you are in. Salvias, for the most part, are called sages. A sage in Northern Arizona usually means Big Sage. It isn't related to Salvias. Here the Rubber Rabbitbush gets called sage. It isn't related to either of the others. Sage is a very general term. It is kind of like labeling something a "live oak". Your live oak in Oklahoma is different than a live oak in California. A live oak isn't a particular variety; it just means an oak that pretty much holds its leaves year round. Now calling something a dead oak means something very specific, but that is another story. We could get into a discussion about the botanical names of the different sages but I don't think that would help you figure which ones would work for you. Instead here is a list of some of the plants that work. Call them whatever you want.

Salvias- There are more that 300 varieties of salvias in the world. The culinary sage is a salvia. Quite a few of them are adapted for our southwestern climate and conditions. In fact I would say they are some of the best and easiest shrubs to grow. Most like our soil, have colorful flowers, attract hummingbirds and are very drought tolerant. Sizes range from 2'x2' for Mexican Blue Salvia to 5'x5' for Chaparral Sage. At 3'x3', Autumn Sage is probably the longest-flowering and easiest to care for shrub there is. It is native to Texas and Mexico. Locally we have Salvia Lemmoni from the higher elevations. It works well in partial shade. Also there is Salvia Parryi with light blue flowers. Unfortunately, this one kind of smells like a cat litter box so we don't grow it too much any more.

Artemisias - This group of sages is probably best represented by Big Sage (Artemisia tridentata) Big Sage grows in northern Arizona in Apache, Navaho, and Coconino counties. It can get fairly large (7'x7') and grows with pinion and juniper or in pure stands. It is gray. This particular plant demands good drainage. Over water it and it will die. The flowers of Big Sage are collected by some hippie-types for burning as incense. Personally, with the wildfires the last couple of years I believe I have smelled enough burning vegetation. Locally we have White Sage (Artemisia ludovicana). This plant grows mostly on oak covered hillsides. It is very tough and can take almost any conditions. It gets about 3'x3', spreads rapidly and is almost indestructible. Like many other Artemisias, the flower is almost the same color as the plant. This is very different than the Salvias.

Rabbitbush- Rubber Rabbitbush is a gray plant that grows along many of our washes. It can get almost 6' tall. It is another plant that requires good drainage. I like this one because it flowers in the fall after almost everything else is done. For that reason it gets a lot of attention from the butterflies.

I guess the word sage is a lot like the word chili. (We are talking about the food dish not the pepper.) If you have nothing better to do, you can spend your time arguing about exactly what it means. My vote goes for just enjoying them.

Field Guide to Local Roadside Wildflowers

I've been doing a fair bit of driving lately. Considering the amount of moisture this winter there is a decent wildflower show going on. These wildflowers can be a little hard to ID at 55 mph or more. Of course we are all in too much of a hurry to stop and actually look at them. So as a public
service, here is the "**Incomplete Field Guide to Roadside Wildflowers in Southeast Arizona**". Not much information in it but it is easy to use.

WHITE FLOWERS- It has been a good year for White Evening Primrose (Oenothera caestiposa). This low growing primrose grows on our hillsides. It has white, often multiple flowers that open in the evening and close up in the heat of day. The flowers are up to 3" across. The White Evening Primrose is perennial and grows back every year from its fleshy roots.
Similar species: This primrose can easily be confused with plastic grocery sacks that have blown out of someone's pick up. Remember primrose blooms at night and morning so if you see a white patch in the middle of the day it is probably a grocery sack.

PRICKLY POPPIES - This flower is 2'-3' tall and is in bloom now by the side of the road. It probably likes the extra run-off there. The 2"-3" flowers are white with a yellow- orange center. The stems are a blue-green color and covered with spines, as are the leaves. Personally, I'd rather handle cactus.
Similar species: There are two species of Prickly Poppies in Southeast Arizona. It's not necessary to tell them apart because they both are spiny, invasive, and a bear to get rid of once you have them.

PINK FLOWERS- Parry's Penstemon (Penstemon parryi) has 3' spikes of hot pink flowers and is in full bloom. It is one of the most drought tolerant penstemons and one of the first to bloom. Early arriving hummingbirds are attracted to it. It is a perennial that does a good job of self-seeding.
Similar species: Superb Penstemon (Penstemon suberbus) looks very similar, especially at 55 mph. It also flowers basically at the same time. However, it doesn't occur in Santa Cruz County and is much less common than Parry's. If you were betting, put your money that you are looking at Parry's.

CALLIANDRA (Calliandra eriophyla) is a low growing (2') spreading shrub. Its pinkish flowers are feathery and its leaves resemble a mini mesquite. It usually grows on rocky hillsides in clusters. Another common name for Calliandra is False Mesquite.
Similar species: Our beloved Catclaw (Mimosa disocarpa) is sometimes confused with Calliandra. Mimosa is larger when mature

and has wicked hooked thorns on its branches. Calliandra is thornless. Mimosa flowers in the summer with spikes of purple to white flowers.

ORANGE FLOWERS- Orange Globe Mallow (Spaeralcea sp.) can get to almost 4' tall. It is covered with 1" orange flowers. The leaves have 3 lobes with the center one usually being the longest. The leaves are covered with what technical-minded botanists would refer to as "little hairy bumps". In some of our bad drought years this mallow was the only plant flowering this time of year.

Similar species: None, but the Orange Globe Mallow can also flower in white, lavender, and coral.

CALIFORNIA POPPY (Eschscholtzia californica) has orange 2" flowers on bluegreen carrot-like foliage. Most places it grows about 1' tall. The neat thing about this plant is that it waits for the right conditions for a big show. About once every 8-12 years we get enough rain for the big bloom. Until then it waits and blooms along roadsides where it takes advantage of extra warmth and moisture. Usually it's an annual but can summer over if the conditions are right.

Similar species: Mexican Poppy (Eschscholtzia mexicana) is very similar as a seedling. However, it matures smaller and isn't as vibrant in color as the California version. On the positive side, it is much tougher and tolerates drought and heat better.

PURPLE FLOWERS - Loco Weed (Astragalus sp.) lies extremely flat along the roadside. It has small pea type flowers and gray green foliage. In drier years it grows a couple of inches across. This year most plants are about a foot or more. It is an annual and will dry up and go away once it gets hot.

Similar species: One of our native verbenas (Glandularia bipinnatifida) is looking good especially in the burned area. It grows about 6"-1' tall and has clusters of purple to pink flowers and green leaves. I've seen this plant in bloom in January. This perennial is good for attracting butterflies and bees.

That wraps up our April Wild Flower Guide. Oh yeah, I just remembered we didn't cover yellow flowers. Yellow is a pretty popular flower color in arid areas. Guess this calls for a follow up article when it gets hot. Stay tuned for the "Incomplete Field Guide to Local Roadside Wildflowers Between Here and the White Mountains".

A Quick Comparison

Generally, I hate to read articles about spring planting. They are so full of upbeat, perky, optimistic information that really has no relevance to us here in Southeast Arizona. I don't need features on "How to Grow Bigger Earlier Blooming Roses". I just need to keep stuff alive until hopefully the summer rains come. That is, unless you are planting a garden.

Gardening and landscaping are two different animals here. Let's go over a few comparisons.

Soil preparation is the most critical aspect of planting a garden. Getting your garden ready involves "improving" your soil by adding organic material. This will lower the pH and improve the fertility. It should improve the drainage as well for most of us. The best way to improve your soil is with a compost product. Manure, if well broken down, works well, too, but it can also raise the salt level. Amending the soil is something that is done every year to replace the nutrients that were used up the year before. For landscaping, go with the ground that you have. Amending is a long and expensive process that usually doesn't work on a large scale.

Plant selection ranks number one for landscaping. Pick the plants that grow here and need little help. They are adapted to our conditions and will do the best over the long haul. You will not have to make up for specific deficiencies that exotic plants crave like humidity, soil fertility, or just plain water. Garden plant selection is easy. Pick what you want to grow and eat. Remember what has done well in the past and ask others what has worked for them. One word of warning though, every year is different so don't get stick in a rut.

Watering is different for both landscapes and gardens, too. Obviously you water to keep both alive and make the plants grow. But in the landscape watering is done deeper and less frequently. This encourages the root system to develop and spread. Ideally, over time, these plants can almost be weaned off irrigation altogether. Treat your garden plants like they are rock stars. Water them frequently, feed them if you want, and push them to produce right now. Get what we can out of them because they are only supposed to produce for a few months anyway. Most are history with the first frost.

Weed control is common ground. Weeds are bad news for either situation. I experimented in my garden last year with weed barrier. This product looks like shade cloth. You put it on the ground, weight it down so it doesn't blow away, and cut holes in it where you want to plant. I don't believe I had to pull a single weed last year. It really works. It is not the prettiest thing in the world (my wife said it looked antiseptic) but by the end of the year it was completely covered by plants and you couldn't see it anyway. I am definitely going to use it again this year. Do not use black plastic for the same purpose. Weed barrier is a woven material. It lets water and air through so the ground can breathe. The plastic doesn't let water in, can superheat the ground, and bake roots. Black plastic is bad news.

Hopefully this was some information that you can put to use. It is still nice outside, so stop reading those optimistic spring planting articles and go outside and get some work done.

A Deep Philosophical Discussion

In almost every column I have written about plants or plant selection, the emphasis is on using native-type species. It is simple; we use them because they work in our conditions. However, there are plants out there from other parts of the world that meet our criteria and are what would technically be called "good plants" to use. These are drought tolerant, wind hardy, heat and cold proof, wildlife attractants, and non-invasive. It would be wrong to ignore them and not write about them for two reasons. First of all, they do work and add some good options for plant selection in our area. Secondly, and more importantly, if every column was just about native plants at some point readers would say to themselves "Oh crap, not another article about native plants. I think I'll skip this column and just see if there are any Reality Shows on TV."

Sometimes a plant is just a plant. Native or not. Here is a true story about a passion vine and how we can get too hung up on origin and not pay enough attention to its performance.

About 10 years ago a friend who worked for the Dept. of Agriculture brought me some seed from a "native" passion vine, that was given to her, which had been collected on a hike in the Mule Mountains near Bisbee. The plant was growing wild and I was told that it had purple and white flowers. I found somebody to grow it out and looked up the description in "Arizona Flora", which is the definitive listing of all plants known to grow in Arizona. The book is a bit technical for me but I didn't think this was too hard. Purple and white flower, found in Southern Arizona and Mexico. It has got to be Passiflora bryonioides. We had captured a native species, and better yet, this species was known to attract the Western Gulf Fritillary Butterfly! This was big. It was time to get this baby in production.

At first, we couldn't keep up with demand. As soon as a crop was rooted, it was sold. I tried one at home to test it out. It grew like a weed and kept its leaves in single digit temperatures. The flowers, of which there were more that 50 on a 15' plant, were beautiful. They were about two and a half inches in diameter with white petals and a purple center. And the plant did attract the Western Gulf Fritillary Butterfly. It grew in sun or part shade. Overall it was nearly the perfect plant. The word spread and soon far away outposts like the Sierra Vista Garden Club and Tohono Chul Park in Tucson were offering the plant. Unfortunately for me, but fortunately for good science, the word may have spread a little too far.

One day I got a call from a guy in Rochester, New York (sorry I can't remember his name). There was a massive series of books being published which would cover ALL the plants in North America. This fellow had been assigned to write about the native passion vines. He had heard that I had Passiflora bryonioides. Could I send him a plant so he could study it for his research? Like a proud papa I sent him a plant. A week later the phone rang and the disappointed person on the other end told me "that's not bryonioides, it is cearulea which is a very common species from South America. It can be transferred to the wild by birds that have eaten the seeds. It has been grown in the nursery trade for years". In other words he was politely telling me, thanks for nothing.

A funny thing happened at the nursery level then. When I stopped telling folks the plant was "native" the interest went way down. Soon the number of sales fell off and several people that were producing this plant on a local level stopped. This passion vine was now as welcome as a hornworm in the tomato patch.

So what do we make of all this? Isn't it the same plant that everyone thought was so wonderful? After some deep thought I reached the obvious conclusion: if we had just shot all the birds in the Bisbee area, none of this misunderstanding would have ever happened. Oh yeah, and maybe there are times when a plant needs to be judged on its own merits and not its origin.

Non-Natives That Make the Cut

Last time out we covered why we shouldn't get caught in the "if it ain't native it ain't squat" rut. Here is a short list of imports that have been proven to work well here and meet our list of being "worthy" plants. Each of these plants is drought tolerant, colorful, non-invasive, and most attract our native wildlife. The list goes from smallest to largest.

Purple Cup Flower is only about a foot tall and wide. It tolerates almost any exposure between full sun and full shade. From about Mother's Day on it is covered in one inch purple flowers that attract butterflies. Although the plants aren't particularly long lived, they should last for several years and spread themselves by seed. It is native to Argentina.

Jupiter's Beard is another good butterfly plant which comes from the Mediterranean region. Hummingbirds like it too. It can get almost 3'x3' and is topped with balls of pink flowers in the spring and fall. In watered situations it can spread rapidly from seed. I have seen them make it through 0 degrees and come back from their fleshy root with no problem in the spring. Again, this plant tolerates a wide variety of exposures.

Alright, let's just say it. Tombstone Rose doesn't come from Tombstone, it comes from China. No big surprise there as just about everything else I've seen in Tombstone comes from China, too. It still is one of the best vines to cover a hot south or west facing wall. It works in places where most other vines would burn up. Tombstone Rose: the plant too tough to fry.

Russian Sage is one of the best plants to be introduced in recent times. It gets about 3'x3' and has spikes of purple flowers from spring to the first hard frost. This bloom attracts butterflies and bees. Russian Sage really looks "native" and works well as a transition plant from the wild to cultivated areas. Most critters like deer and rabbits leave it alone. Of course, Russian Sage hails from (you guessed it) Pakistan, Afghanistan, and a few other "stans".

Our roll continues here with Butterfly Bush, another plant from China, and also Taiwan. Butterfly Bush can get kind of big (10'x10') so it doesn't know if it is a large shrub or small tree. I guess it is in how it is pruned. One fall I tried to count the butterflies on a plant at the Elgin School. I got to about 150 on a major branch and quit. In addition to the traditional purple flowers, Butterfly Bush now is available with white or gold blossoms.

From the Philippines, Taiwan, and China comes the Chinese Pistachio. This is a fairly large tree that can provide dense shade. The foliage turns a bronze red to pink in the fall. It is probably the most consistent performer for fall color here and contrasts nicely with the yellow and gold of the Arizona Ash. This isn't the pistachio that grows the nuts, it is a relative. This tree however is often used as the root stock for the fruiting variety. That should tell you something about how hardy it is.

Use It Twice

Everyone has a few guiding philosophies to live by. Some are as broad as the "Golden Rule" or as simple as "if it ain't broke don't fix it". Personally I like the "ain't broke" philosophy but I also believe in the "if it is broke don't fix it either, just put it up in the pole barn" school of thought. Sound reasoning like that has gotten me where I am today. When it comes to landscape related issues my philosophy is "use it twice."

Using it twice is just a way to double your return on effort and water. A Redbud (Cercis occidentalis) gets used once. It is beautiful in spring when it is covered in magenta pea-shaped flowers. Then it leafs out and cutter bees decimate the leaves. Everybody loves them for about two weeks a year. On the other, hand a peach blooms with pink flowers in the spring. It takes about the same amount of water and room as a Redbud. After flowering, if it doesn't freeze or the wind doesn't blow too hard, the peach tree does something the Redbud can only dream of. It has peaches. Two good uses from one tree.

Use shrubs twice, too. Instead of something with a short flowering season or the old standby low growing junipers, try plants with a long blooming period that attract some kind of wildlife. Autumn Sage, Blue Salvia, Lavender Spice, and Russian Sage all bloom for much of the growing season, like our soil, and attract hummingbirds and butterflies.

Bedding plants work this way also. The other day we watched at least 3 different species of butterflies land on a mix of Coreopsis and Gaillardias (Indian Blanket Flower) in less than a minute. Snapdragons, Pentas, Cosmos, and Zinnias will attract desirable wildlife. All thrive with only minimal care.

Water is definitely something that should be used twice whenever possible. State law now allows the use of gray water. "Used" water from washing machines and showers can be run to irrigate trees, lawns, or gardens. This is good water that would just be flat out wasted if it wasn't used twice. No special treatment is necessary to reuse it, just run it out to where you need it. One of the best gardens we ever had was watered entirely by a washing machine. Every time we did laundry, it got watered. No need to even wash the vegetables off before we ate them.

Jim's (Water) World

After a long, hot day's work, there is nothing more refreshing than sitting around the old stock tank and watching the goldfish swim. This little bit of happiness can be achieved far easier and cheaper than you might think. In order to get you started, here is a complete guide to your very own water feature.

In our world, the water feature is a closed system; it basically takes care of itself. Just add water. Before I get started on putting this feature together, I should probably put in a slight disclaimer. If you are someone who needs the water to be crystal clear at all times, this system is not going to work for you. Of course, if you are someone that demands everything to be crystal clear at all times, go get help. Life just ain't like that in this world or any other. Now that's covered, let's get started.

Step 1. Start by rounding up all those catalogs, articles, and printouts about ponds that you have been collecting. Organize them neatly into a pile. Now, toss them all out. Ninety-nine percent of what they are trying to sell you is unnecessary junk. While discussing how to set up water features, an old friend asked "You mean I just spent $16 on a bunch of useless crap?" Yup.

Step 2. Pick out your pond. I started about 12 years ago with a stock tank that was left over after we sold off some horses. That one was about 6' across and 2'deep. You can use old bathtubs, stock tanks, wash tubs, or anything that will hold water. Size doesn't really matter, but it should be at least 1' deep. I have noticed that whatever size you start with, you are going to wish it were a little bigger. If the selected vessel leaks, a little aquarium sealer or silicon caulk should fix it.

The tank should be placed where it gets at least half a day of sunlight. This will help bring out the best in the flowering plants. If you want a more natural look, bury the tank (except for about 6") and line the edge with rocks. Through a great deal of scientific experimentation, I have learned that it is easier to move a tank before it has been buried and filled with water.

Step 3. Turn on the hose and fill 'er up. For those using well water, that is all there is to it. If there is chlorine in your water, it needs to come out. Letting the water stand for a few days should do the trick. If you can't stand the wait, go to an aquarium store and buy some dechlorinator. I hope you have been following this set of complex instructions because this is where it starts to get interesting.

Step 4. Pitch in a couple of shovels of dirt or manure into the water. The dirt or manure will act just like sourdough starter does in bread making. It jumpstarts the nutrient building process that plants need to be healthy. After a couple of days the "starter kit" should settle to the bottom.

The man who I learned this approach from told me the worst thing you can do to a pond was to clean it. All the nutrients are lost. While he was telling me this, I remember thinking two thoughts. "No cleaning? This sounds like my kind of method." Which was followed by, "I'll have to go shovel the gunk back into the tank that I cleared out this morning before I went to town."

The Continuation of Jim's (Water) World

OK, we left you with a tank filled with non-chlorinated, nutrient-enhanced water. Now it is time to finish it up.
Step 5. Time to plant. Water plants are kind of the opposite of land plants. It is all you can do to grow land plants and it is all you can do to contain water plants. There is a reason why water plants are more expensive than land plants. The person selling them knows that they can only get you one time.

Water plants occupy one of 4 zones in your tank. Use some of each so that the water feature has a more natural look to it. Floaters do exactly that; they stay on the surface and really help filter the water. My favorite floater was water hyacinth. I say "was" because they have been banned in this state as being invasive. Now I use water lettuce in their place. If you pick up one of these plants, you will see a long stringy root system. These roots both filter your water and provide a hiding place for the fish.

Bog plants are the plants that would grow around the edge of a pond or lake naturally. They may be submerged for part of the year. In your water feature bog plants should have their containers placed just under the surface of the water. If you are using a 2' deep tank, a cinder block put on its end works about right to get them to a good level. Many bog plants flower and can add some color. I like Pickerel Weed (purple), Cardinal Flower (deep red), and Obedient Plant (pink). There is also a medium sized Cattail (3'-4') that looks natural in water features. Always put your Cattails in containers or they can take over your water world.
Aerators are the easiest plants to deal with. These plants are basically seaweed. They don't do much visually but do provide some extra oxygen for the fish. Just pitch a handful in when you are getting started and forget about them. They don't need to be planted.

Finally, we have the fourth group of plants which are those that are put in containers and placed on the bottom. The most important of these are the water lilies. Most water lilies like to be in 2'- 4' of water, although there are some mini lilies that do well in shallower situations. They need at least a half day of sunlight to get the best blooming effect. The hardy ones mostly bloom in white, yellow, or pink. The tropical varieties have exotic shades of purples, reds, and other brilliant colors. Unfortunately, they are very expensive and will not survive the winter here. Don't ask "what if I bring them indoors for the winter?" because you will probably not get around to it and they will die leaving you bitter about pond plants. Stick with the hardy ones.

Don't skimp on the plants. A water feature without plants is just a stock tank or bathtub. They keep the water relatively clean. Also, keeping the tank at least 60% - 70% covered with plants is the best defense against algae.

Once a year, or more likely, whenever you get around to it, it is time to separate and repot the plants. This is not a delicate process and you really can hack them into several pieces and repot them. Don't use potting soil to replant them. It is full of organic material and will probably just end up floating to the surface. Instead use a heavy clay soil. Water plants get their food from the water not the soil.

Step 6. Time to go fishing. Fish are the final piece to the perfect (almost) water world. I have seen tanks with goldfish, Koi, bass, and catfish. When you are getting started however, always start with what are referred to as feeder gold fish. These are available from local pet stores and should cost about $1 for 10. Should there be a problem, you will be out only a buck. Goldfish take the cold really well. I have had several inches of ice on a tank in the winter with no problems for the fish. Always give them shelter like rocks, plants, or the cinder blocks we talked about before. The shelter may save them in case a Great Blue Heron comes over for lunch. The fish help add fertility to the water. Also they eat the mosquito larvae. Never feed your fish! There is plenty for them to eat in the pond. Besides, these are supposed to be "working fish", not pets. The next time you are in a situation where the local ranchers are sitting around drinking coffee and bragging on their cattle dogs, pull up a chair and join in with a story about your "working" fish. I guarantee they will take notice.

Step 7. Be a little patient. It takes a little time for a pond to reach a state where it is happy. It may be necessary to tweak it a little by adding a few more plants or taking out a couple of fish. My best tank is about 10 years old. It has mostly water lilies in it and the water is clear year round. I just have to add water to it once in a while.

Well that is about it for the perfect water world. If only that other world were as easy to deal with.

L.E.M-ing

We don't hunt rabbits with a cannon and most of us don't drive finishing nails with a sledgehammer. Why then do we run out, buy and apply heavy-duty pesticides every time a leaf turns color. Every year I get asked "my plant has a brown tip on it, what should I spray it with?" I usually try and get the person to slow down and figure out what may have caused the damage, does it really hurt the plant, and will spraying do any good. I am a big believer in using the Lightest Effective Method (L.E.M.) that will get the job done. You can call it whatever you want, but I hoped that if I coined a new age yuppie-sounding acronym I might really make the big bucks. Here are a few L.E.M. tricks that work.

Water. You are probably tired of hearing this but water is the most important factor in the health of a plant. Insects will prey on stressed plants before they go after healthy ones. Keep everything well watered.

Water can also be used to get rid of some pests by simply blasting them off with a hose. Red Spider Mites can often be handled like this. You might have to repeat this technique a few times for it to really work.

The L.E.M. approach to soft bodied insects like aphids and thrips would be soapy water. Put about a tablespoon of liquid soap in a quart spray bottle and let them have it. The soap coats their bodies and they die. (Kind of sounds like an old hippie joke.) This soap mix is basically the same as the "Safer Spray" that is sold in organic gardening catalogs for a lot more money.

Try using your soap spray on grasshoppers or beetles and you will end up with ... clean grasshoppers and beetles. To deal with these kinds of pests you have to pick it up a notch. In my garden, I keep a jar with alcohol (rubbing, of course) on a cotton ball. Last year I was able to easily control the beetles on my green beans and squash bugs by picking them and tossing them in the killing jar. It's a good idea to do this at least once a day. This approach might not work in a commercial setting, but I'm not about to spray raid on anything I'm going to eat.

Grasshoppers are another matter. When they are a problem there's usually too many to hand pick. Semaspore is a good way to start control. It is a grain bait that when eaten, shuts down the digestive system of grasshoppers and crickets. It has no effect on anything else including the birds or lizards that eat them. You have to apply Semaspore when you first start seeing the grasshoppers. It usually kills about half the ones that consume it.

Shutting down the digestive system is also a good way to control Tomato Hornworm and Grape Leaf Skeletonizer. In fact, any caterpillar can be controlled by using a product called Bt (Bacillus thuringiensis). Bt shuts down the digestive system causing them to turn black and die. Like Semaspore, it does no harm to anything that might go around eating dead, black caterpillars.

Everyone has their own tricks and home remedies to deal with pests. Let's not lose sight of the point that an approach needs to be easy, low cost, and most important, it needs to work to be a part of the L.E.M. system. There are times when herbicides or pesticides are the best solution. If you are trying to eliminate Bermuda grass, Roundup or other glyphosate is the lightest, effective method. Go ahead start L.E.M.-ing. Bumper stickers available soon.

Just Nuke 'Em

A while back I wrote an article that suggested dealing with pests with the lightest effective method (L.E.M.ing). It basically made the point that you don't need to use a chemical type solution on a pest if there are other less toxic, cheaper methods that work. The key phrase here is "that work". In the case of gophers, never mind all the organic, sonic, live and let live mumbo jumbo. Just nuke 'em.

I saw the telltale signs the other day. There was a new small mound of freshly piled up dirt and about a 2" hole a ways away from it. The gophers are coming, the gophers are coming! Let's backtrack for a second. These are pocket gophers not moles. To the best of my knowledge we don't have moles in our part of the country. Pocket gophers live in underground tunnels. I've been told that a single male gopher can have a tunnel network of about 200 foot square. The mounds that you see are vent holes that aerate the tunnels. Most of the tunnels are about 4" to 6" underground. Gophers definitely prefer softer soils, you probably won't see them in clay or rock.

If gophers would just dig their tunnels and vents we would leave them alone. We might even give them credit for aerating the soil. But here is the problem with these varmints, they eat roots. They eat them faster than a plant can grow them. I have had some fruit trees in the ground for 5 – 7 years that didn't look much different from the day I planted them. Of course since I started nuking the dang gophers, they are looking better.

There are a number of ways to deal with these critters. A lot of catalogs advertise a device that creates a sonic noise that drives them away. Research has shown that the only thing that goes away is your hard earned money. The sonic method doesn't work. Trapping with special gopher traps works but takes practice and patience. I thought I might try a "natural control" years ago. I caught about a 4' gopher snake and let it loose down a gopher hole. I got up the next morning as it was getting light, grabbed a cup of coffee and headed down to the orchard. Unfortunately, I had completely forgotten about the snake which was lying coiled up in the grass. I had to jump out of the way to avoid stepping on him. By the time I returned to earth my brain had processed that it was a harmless snake that had just caused me to jump and holler. Less than a week later I saw some new mounds. So much for the natural approach, it was time to just nuke 'em.

The best way to get the job done is with a grain bait. The smoke bomb control (fumigation) doesn't seem to work in our soils. With the bait, the gopher eats the poison and dies underground. This is good for a couple of reasons. Dead gophers make lousy looking yard art. Also when they die underground there is little danger of dogs, birds, etc eating a poisoned animal. Statistically, they would have to eat a whole wagon load to have any detrimental effects but why take chances?

The real trick with the bait is getting it down in the tunnel where it can be effective. The directions on the can will tell you to 1) locate the tunnel, 2) dig down to the tunnel, 3) put the bait on both sides of the excavation, to insure that some of the bait is on the side that the gopher is on. I'm sure that this works and I would never tell someone to go against what the directions say but I usually have something else to do besides figure out if the gopher is coming or going. So, here is my own approach, for entertainment purposes only.

Find the vent hole. Run some water down it to make sure that it flows. Don't get carried away and try and flood them out. It will never work. Put the recommended amount of bait down the hole. Run some more water down to flush it down to the main tunnel. Make sure there is no bait left on the surface, that could be bad. Another problem solved. Too bad grasshoppers aren't this easy.

A Field Guide to What is Eating Your Plants

You can find a guide to almost anything these days. There are books for trees, wild flowers, native grasses, and shrubs. It's usually not too hard to identify a plant. It's a little harder to ID what's been eating those plants. Anyone who has planted anything this year has had some casualties. If you have tried to plant without fencing it first you have probably given up by now. If so, skip this column and spend your time looking at the "vacation property" section of the classified.

First, some good news. Because of the drought, there has been very little insect damage so far. I've only heard one grasshopper complaint. Didn't see any hornworms until last week. Nobody has reported the swarms of Blister Beetles that usually come and decimate a garden this time of year. If you have damaged plants in your garden it's probably not bugs, yet. Here's a field guide to what it might be.

Symptom: Seeds don't come up, or come up and disappear.
Possible cause: This sounds like bird damage, especially Curve-bills or Jays. They will eat seedlings as quickly as they come up. Sometimes they even dig in the ground and eat the seeds. You can wait until we have some moisture and replant, or try and protect the area with screening.

Symptom: Plants get smaller overnight, usually in one area at a time.
Possible cause: Mice are probably at work here. I watched my strawberry patch slowly disappear this spring. They ate the leaves first, then the fruit. I set several different kinds of traps before catching any mice. The old-fashioned spring type worked best for me. Use cheese or peanut butter and crackers for bait. Remember though that most things eating your garden are night robbers. You should get out early to check your traps and spring them or you will catch some unintended critters like birds. Those Copper Trogons are really spectacular up close though (just kidding).

Symptom: Up to pencil-sized stem cut diagonally. No trace left of cut-off piece.
Possible cause: The diagonal cut is a telltale sign of a pack rat. I don't know how to get rid of these pests without poison. They can clean off the bait from a Havahart or rat trap without getting caught. I believe they are smarter than I am.

Symptom: Stems cut off at 90-degree angles.
Possible cause: This kind of cut can be rabbit, squirrel or deer. Squirrels can climb most fences or walls. Rabbits can fit through any opening bigger than 2 inches. This includes chain link fencing. Use a Havahart trap to catch them and then get rid of them.
The common thread in most of these situations is not the search for food but also moisture. Many rodents get the majority of their moisture through what they eat. In this case they are drinking your garden plants. Fencing doesn't eliminate the problem. Putting out water for the rats doesn't seem to be a great idea either. Our best hope is (I know you have heard this before) rain and lots of it. Then we can get back to what we should be doing. Complaining about bugs.

TOO HOT FOR COMPANY

PLANTING DATES   p. 140

DORMANCY RETURNS   p. 141

A QUICK BREAK   p. 142

A GOOD PLANT, A BAD NAME   p. 144

PICKING MATERS   p. 146

THE OLD TOMATO BACKSTEP   p. 147

HOW HOT IS YOUR CHILE?   p. 149

WATERING EXISTING PLANTS   p. 151

WATER RULES   p. 152

GOOD QUESTIONS   p. 154

THE FUTURE, PART I   p. 155

THE FUTUER, PART II   p. 157

GOOD BYE OLD PAL   p. 159

MENTAL HEAT RASH   p. 161

EPILOGUE: ADIOS   p. 163

Planting Dates

First of all, no complaining about the rain, clouds, or weather. I don't care if it messes up your work, weekend plans, or made your roof leak. We need the moisture. This is the way we used to get winter moisture back in the good old days, say 10 years ago. Of course complaining that it isn't enough is always acceptable.

Several hundred years ago Ben Franklin coined the phrase, in a letter to a friend, that the only certainties in life were "death and taxes". He had two thirds of it correct. What he should have said was "death, taxes, and people trying to plant tomatoes too dang early in higher arid regions of the Southwest." I will flat out guarantee you that there will be a warm spell this February. Some people will start to get itchy fingers. First we will get a warm and sunny week to ten days. The growth tips on fruit tips will green up and people will wonder if they are going to bloom. Then some knowledgeable folks will get the feeling that "it's over". After that a six pack of tomatoes will be purchased at a large box type hardware store in the low desert. They will be planted here. Finally you get to repeat steps three and four two weeks later when everything freezes. Remember last January? It was much warmer that February and early March.

The big question is when is it a good time to plant? Other parts of the country have a planting date. Way up in the frozen north I am told they plant on Memorial Day. As you go farther south of course the date gets earlier. A "planting date" doesn't work here because our weather is too variable. The lay of the land (microclimates) has a lot to do with how the cold settles. Also having a date set in stone is just kind of boring. It is more fun to play the odds. Here is some inside information to help you place your bets. The numbers come from the Western Regional Climate Center. Being that slight changes in location, elevation, and exposure really change frost patterns, I am using two different examples to help find a good planting date. You can figure which one you are closer to, or average the two.

Sierra Vista is warmer than most places in Eastern Santa Cruz or Western Cochise County. Plant early here and there is a 90% chance of freezing after February 20th. The big decline in risk is from the last week in March to the first week in April. You reduce your chances from 50% to 10% of freezing. For you statisticians out there these numbers are based on 21 years of observation, from 1983 to 2003. I'd be a little more comfortable if they were based on a little longer time period like 1003 to 2003.

The Canelo Reporting station is Siberia by comparison. There still is a 90% chance of frost after April 15th. It still frosts 60% of the time after May 1st. Wait until the middle of the month and the problem is reduced to about 10 %. The latest recorded freeze there is June 6th.

Even if you do duck a killing frost by planting early, is it really worth it? The ground is still going to be cold. Chances are that the real growth won't start until the summer rains. Don't be in such a hurry; there is no prize at the County Fair for the "earliest garden". Besides this is Southeast Arizona. Never do today what should be put off 'til tomorrow.

Dormancy Returns

Quit whining. It really hasn't gotten hot yet. So far this has been one of the most pleasant spring and summer combinations in recent memory. Still, things are drying out. The grasses that were greened up in April turned brown in May. Now they are heading for a scorched shade of bleached tan. The fact that most of us had no measurable precipitation in May has a lot to do with that. But heck, this is what the range is supposed to look like this time of year.

Take a minute and look closely at native trees like the ashes and the oaks. They are green but have stopped growing. Remember the spring wildflowers? They have gone to seed by now. Even landscape trees for the most part have little new growth on them. It is hot, dry, and windy and these are not favorable conditions to promote new growth. We are in a second period of dormancy for many of our plants. Of course the winter dormant time is much more obvious, but right now plants are just maintaining and waiting for the summer rains.

Now if I was reading this instead of writing it I would probably be thinking so what, who cares, and how long until the big wrestling match comes on TV. However there is a practical reason for recognizing a second dormant period. If an established plant wants to kick back for a month or two, let it. Water it enough to keep it happy, usually about one good soaking a week. Don't feed it. Most plants need very little extra food. If you really have the need to feed, wait until the summer rains. You will get more growth in one week of good summer rains than in a month of trying to force things now.

 For plants, summer rains is when the good stuff happens. They aren't real fond of the famous "dry heat". When the humidity goes way up it is great for plants and not so much fun for us. When it is hot and dry plants can give off moisture through their leaves faster than they can take it in. That is why some plants wilt even though they have been watered. Humidity helps to slow down the evaporation rate.
We all know that rainwater beats the heck out of groundwater. Rainwater contains nitrogen. Nitrogen promotes green growth and is sorely lacking in our arid soils. Every time it rains it is like feeding your plants. Higher humidity and free fertilizer means the plants wake up and boogie.

Right now we are already looking toward Mexico to see if there are any promising clouds. Our historical average for summer rains is about 8" to 9" inches for July through September. For 8 – 12 weeks the grass again turns green. We can pull a chair up at night, watch the storms, and try to guess "if that storm is going to hit". Life is good. Then someone will start to complain about the mosquitoes.

A Quick Break

Two questions dominated the conversation at the nursery this week. One was obviously about the good looking clouds and the little bit of moisture that fell from them. Was this an early start to the summer rains and will it continue? Danged if I (or anyone else) knows, but if I had to guess I would say that this was a freak deal and get ready for at least 3 – 4 weeks of the "dry heat" that we all know and love. Let's not forget that we do live in Southern Arizona. There is no easy path between here and moisture.

The second question is easier to wrap our arms around. What is eating the leaves of our roses, peach trees, Red Buds, and anything else that has a fairly broad leaf? That is a simple one to answer, nothing. Nothing is "eating" the leaves. The neat semi-circular cuts taken out of the foliage are being removed by Leafcutter Bees. But they don't eat them, they roll them up between their little bee legs, fly back home, and line their nests with them.

Leafcutter Bees are solitary bees; they don't live in colonies like Honeybees but instead inhabit small holes left in dead branches or trees by boring beetles. The rolled up leaves slide in the quarter inch hole like a trash can liner. Most Leafcutters are about one half inch long and have grey and black stripes on their abdomen.

The question most asked this week is what should be done about them. Again, the answer is a loud nothing. Leafcutter Bees are excellent pollinators. In fact most of the native bees make the imported Honeybees or African Bees look like sharecroppers when it comes to the job of pollination. The Leafcutter Bee that takes a part of the leaf on your peach tree this summer might be the one that makes the peaches next spring. Another of their specialties is sunflowers.

I will take this a step further. It is my belief, completely un-backed by any credible research, that the cutting of the leaf surface this time of year is beneficial to the plants. It removes the excess surface of the leaf at a time of the year when we are having triple digit temperatures and single digit humidity. Look closely at a leaf that has been severely hit by Leafcutter Bees and it almost resembles a mesquite leaf. This is a leaf form that is meant to survive heat and wind. It is a leaf that has had all unnecessary parts removed. You should be thanking these critters. It should be added that I have never seen or heard of a time when these bees caused permanent damage to a plant.

I recently sat in on a talk about insects by a County Extension Agent (not Santa Cruz or Cochise). This guy was perhaps maybe not real familiar with the workings of higher elevations in the Southwest. When he showed a slide of the Leafcutter Bee I asked him about my theory of them being beneficial. He responded that he didn't believe that removing leaf surface from actively growing plants could possible be beneficial. I politely responded with "oh yeah." Let me translate "oh yeah" for you. Basically it means "Hey Mr. City Slicker, get out of your air-conditioned office once in awhile and you will see that when it has gotten hot and dry in arid lands, plants want to go dormant again to save energy until there is some summer moisture. That is unless they are being unnaturally pushed by lots of moisture and food." End of translation.

Maybe I am being too hard on this fellow. After all I am just the local nursery guy. It is not like I have a lot initials after my name like PhD or CEO like he does. Perhaps if I did I would have more instant credibility. Wait, let me check my shorts. Good deal. From now on I will say that this was written by James Koweek, BVD.

A Good Plant, A Bad Name

This is worse than a boy named Sue. The Desert Willow (Chilopsis linearis) has long thin leaves so someone thought a handle like "Desert Willow" was a good idea. Calling it a "willow" makes it seem like a water sucking, root invasive pest. The Desert Willow is neither. It is however one of the best, fastest growing, showy flowering trees for our area.

Every year around May the Desert Willow blooms with a profusion (that means a whole lot) of light lavender to purplish orchid-like flowers. This year, despite the lack of moisture, Desert Willows are in full bloom. They seem to prefer the dry canyon bottoms but really aren't picky about soil types. Unlike some other wimpy flowering trees (Redbuds) that flower one time in the spring and are done for the year, Desert Willows can have some repeat blooms. This usually happens after a rain.

Desert Willows can have a light sweet smell to them, which can carries for some distance. The flowers attract loads of hummingbirds and butterflies. After flowering, the tree sets long thin seed pods. The pods hang on the tree until they are dry then split open. I'm told that the small web-like seeds are used by hummingbirds in nest building.

Desert Willows usually grow about 20' high and 15' wide in a landscape situation. In the wild they get much bigger. The current record for the largest one is in Globe, AZ. It stands 68' tall and is 13' around. The Desert Willow is a tree that looks particularly good when planted in clusters to form a grove.

These days almost all nursery stock is propagated by cuttings. This ensures that the flower color stays constant. If you grow them from seed you won't know what color flower you have until it blooms. 10-15 years ago most Desert Willows were on the pale side. Through the years, plants were selected and propagated for darker, purpler blooms. It still helps to see your plant in bloom to know exactly what you are getting. One variety called "Burgundy" has almost a maroon flower. However, "Burgundy" tends to be shrubbier and slower growing than other types.

I have a good number of Desert Willows growing native on my property. This year they have gone through temperatures as low as -5 degrees with no problem. Every morning when I walk by them to feed, I get a whiff of the fragrance. It's one of life's little pleasures while we are waiting for rain.

Picking Maters   5/10/02

(The next two articles show how you can change your opinion on a subject. Put the books and magazines down, experience is still the best teacher.)

This was going to be an article about chiles but most of us have had a hot enough time the last couple of weeks so we'll start with tomatoes instead. Here are some suggestions to help increase your tomato harvest.

Plant them deep. Tomatoes have the ability to root from the same places where they leaf out. Strip off the leaves from the bottom two thirds of the plant and bury it with the top third sticking out of the ground. This gives you a plant that will develop much sturdier stems. Old-timers would often lay the whole plant on its side, underground, with just the tip sticking out. This is called trenching.

Check to see if your plants are determinate or indeterminate. Determinate plants all ripen at about the same time and then are finished. These are good for salsa or canning. Indeterminate tomatoes ripen a few at a time and keep producing until the first frost.

Plant several varieties. Anyone who has been through more than one summer here knows that they are all different. Different conditions, especially the timing of rains, favor different varieties. There is no single "best type" for our area. Four years ago my best variety was Early Girl which won me a Blue Ribbon at the County Fair. Three years ago the grasshoppers ate everything. Two years ago it was a yellow type called Lemon Boy. Last year it was two varieties that I'd never heard of until I planted them, Fourth of July and Abe Lincoln. Planting different varieties will also stretch your harvest season. Early varieties can ripen in as few as 50 days while some of the beefsteaks will take as many as 100.

Try some "heirloom" varieties. Heirlooms are old, pure strains that have been preserved for generations. They are not hybrids. Heirlooms come from all over the country and the world. Brandywine, Abe Lincoln, and Nebraska Wedding are a few of my favorites. OK, I admit I haven't planted the Nebraska Wedding yet but I like the name.

Get acquainted with Bt, a non-chemical pesticide. If you have tomatoes you are going to have tomato hornworms. They are the caterpillar form of one of our sphinx moths and they like to dine on your plants. Bt shuts down their digestive system. They turn black and die. It's a beautiful thing. Bt doesn't affect anything else but caterpillars so it's very safe to use.

Use paste tomatoes for salsa. I've talked with some very good salsa makers lately and most agree on only one thing. They use paste tomatoes (Roma, Amish Paste, etc.) for their salsa. It makes it meatier and less runny. Paste tomatoes usually have fewer seeds, too. If you're making salsa this year try paste tomatoes. Also, remember where you heard these good tips and bring some by.

The Old Tomato Backstep 5/17/05

In the past several years I have written articles singing the praises of heirloom tomatoes. An heirloom tomato is basically a variety that is open pollinated (non hybrid) and is at last 50 years old. Usually the seeds have been preserved by a family, or in a region, for generations. They have great names like Nebraska Wedding, Mortgage Lifter, or Abraham Lincoln. And, unlike hybrids, if you save the seeds, the next crop should be the same as the one that you harvested from. You don't have to buy new seed every year. People claim that they taste like an "old fashioned" tomato. The flavor reminds them of something their grandmother used to grow. Heirlooms have been the rage the last few years. Almost every publication from the organic gardening magazines to the Farmers Almanac has featured articles on them. Martha Stewart probably started some on the window ledge of her lonely prison cell. However, I have learned there is a slight problem with devoting a large portion of your garden to heirloom tomatoes here in SE Arizona, they don't produce worth spit.

The main reason for this is that heirlooms by their nature are regional. People would save the seed from their best tomatoes for planting the next year. After a while that would become a variety that was well suited to the growing conditions of that local area. I know it is not great insight to tell you that our growing conditions have little in common with Amish Country, Germany, or the Midwest which is where many of the heirloom varieties originated. For me the great wake- up happened when I was shipped an heirloom variety called "Quimbaya". When I asked about the origin of the Quimbaya I was told that it was the "legendary heirloom tomato of the Amazon Rain Forest". Well what the Sam Hill do we have in common with the growing conditions of the Amazon Rain Forest?

There is another reason to look beyond heirlooms. Ever since the dawn of agriculture people have sought to improve their livestock or crops by breeding. I don't believe anyone ever said "let's take your stud with the bad legs and breed it to my slow mare and see what happens." Hybridization has always been a way to improve desirable traits. Of course not all breeding makes for better plants. Some has been done for totally useless characteristics to the home grower like appearance or better packing and shipping qualities. However, some hybrid varieties like Early Girl and Lemon Boy were successfully developed for increased flavor, earlier harvesting and greater production.

To be fair, there are some local success stories with heirloom tomatoes. Tillman Calvert told me that he grew a Mortgage Lifter last year that was 16' wide and about 8' tall. He harvested about 150 fruits from it. I believe I'd try a couple of those. Also, I was in a garden in Canelo that was nothing but heirlooms. It was one of the prettier plots that I have ever seen. There were tomatoes from white to orange to deep red and every shade and shape in between. The single best tasting tomato I have had in the last few years was a Pink Caspian that I grew. Unfortunately, the plant only produced two fruits that entire year. I can't justify the space and water needed to grow a mature tomato plant for only a couple of fruits.

I am not going to give up on heirloom tomatoes completely. We just won't depend on them as a food source. They are kind of like buying a lottery ticket when there is a huge jackpot. The odds of it hitting are slim but there is some entertainment value to be had in the waiting and hoping.

How Hot Is Your Chile?

(Jim's note: Since this article was written there has been a new King of the Hot Chiles crowned. The old Red Savina is out and the new champion is Bhut Jolakia. It is of Indian origin and has been scientifically measured at over one million scoville units. That is over three times hotter than the hottest documented habanero. I am sure that is just what the world needs.)

No doubt about it, we can grow chiles with the best of them. In fact, Arizona is the second leading producer of green chiles behind only New Mexico. It is a little know fact that most of the green chiles grown in New Mexico come from a seed source in Southeast Arizona. Our climate here in southeast Arizona has a lot more in common with Deming or Las Cruces than with Tucson or Phoenix.

Peppers have a tremendous range of flavors and heat. The heat of any pepper can be measured by something called Scoville Units. The Scoville Unit's scale runs from 0 (Bell Peppers) to about 300,000 (Red Savina Habanero). For comparison, an average jalapeno runs about 5,000 Scoville Units. That means a habanero could be up to 60 times as hotter than a jalapeno. John White, the Ag. Extension Agent from Las Cruces NM., recently told me that a habanero of 1,000,000 Scoville Units is in the works. How did we ever get along without that one?

Environmental conditions also affect the heat and flavor of chiles. Soil type, amount of sun, and especially moisture all play a role. People from Hatch, NM claim it's the soil that gives their chiles the unique flavor. A well watered, unstressed plant tends to have milder fruit. I've grown Anaheim chiles that were hotter than jalapenos by keeping them dry. Most peppers don't reach full flavor or heat until after they have turned color. A green pepper is an unripe pepper. Friends don't let friends eat green peppers.

Here are a few varieties that are worth trying in our area. Anaheim types (500-1500 S.U.) like Big Jim, Joe E. Parker, etc. do well. Use them in salsa, fresh-roasted, or rellenos. It's not summer if your burger doesn't have a roasted chile on it. Jalapenos (2500-10,000 S.U.) are good fresh or in salsa. They freeze well so you can have them all year.

Anchos have a good flavor fresh-roasted or stuffed. Supposedly they are mild but the ones I grew last year were hot enough to burn you eyes just looking at them.

Of course everyone needs to grow a habanero (80,000-150,000 S.U.) or two. They are beautiful, hot, and deadly if you don't know what is coming. A habanero takes a long season to mature. I put one or two in a pot of beans.

Italian Sweet Peppers are really good fresh or roasted. Not many people are growing them here but I think they should do better than the bells. They definitely are more flavorful than your average green pepper. I'd try the heirloom Corno di Toro or some of the pimiento types.

If you like to eat hot peppers and want some easy money go to the Santa Cruz County Fair. The annual Jalapeno Eating Contest pays good money to whoever eats the most jalapenos in 60 seconds. The winner usually eats about 5-7 so start practicing. Don't worry; the contest is right next to the drink concessions.

Watering Existing Plants

Water. It always comes down to water. Right now good watering is more important than ever. We've gone four and a half months with little to no moisture and it is beginning to warm up. Things are starting to get a little thirsty. Your plant's best friend right now is probably your hose. Their second best friend could be a water probe.

A water probe is a metal stick that can be pushed into the ground to see how far water has penetrated. They should sell for $5 - $10. A piece of rebar works just as well. After you have watered a plant for a while, stick the probe in the ground and see how far the water has gotten. A good general guide for water penetration would be 6"-12" for groundcovers and grasses, 1'-2' for shrubs, and 2'-3' for trees. The amount of time that you have to water to get to these depths will vary depending on soil type. Sandy soil saturates quicker than heavy clay soil, but it also dries out faster.

Where you put the water is as important as how deep it goes. Never water next to the base of an established plant. Research shows that a high percentage of a plant's feeder roots occur in a 4'- 6' zone outside the drip line. The drip line is the pattern that would be created if the outer branches of a plant were traced on the ground. Your hose (or hopefully soaker hose) should be placed in that zone outside the drip line. Remember that less frequent but longer waterings are much better than shallow frequent watering for root development.

The third ingredient in good watering practices of established plantings is mulch. A good 3" - 4" layer of mulch should cover the entire root zone of the plant but shouldn't touch the trunk. Hay, straw, compost, bark, or gravel are all good mulches. They help keep moisture in and keep the roots a little cooler. Personally, I like straw because it's cheap and I feel that the light color reflects some heat. It isn't always possible to mulch a large existing tree but overall, mulching can reduce your waterings by as much as a third.

The real question is how often do I water? The answer is when they need it, which is probably right about now.

Water Rules

One of the more important local news stories broke last week and I am surprised there wasn't more talk about it. The Sierra Vista Herald ran a story under the heading "Bella Vista declares water restriction on south system". The important part of the article is that Bella Vista Water Co. has completely banned ALL outside water use for a portion of their customers. This includes watering of plants, washing vehicles, filling of swimming pools, or any other outside water activity. The first time offense is a warning and they will disconnect you permanently if you are caught again. They have even stopped hooking up new customers in this area.

According to the article and people that I spoke with, the restrictions came about because the demand on the water system, which averages 3000 gallons per house per month, is greater than the rate of recharge in the wells. Bella Vista claims that the problem is a result of the drought coupled with the high demand. You can read what you want into this explanation. I know people who live in this area and they are water conscious. I am betting that either a few houses are burning a lot of water, or that there are more houses on the system than it can handle. The real questions are is this the only place in the Southwest where restrictions are going to happen and what can you do about it?

It is easy to answer the first part of the question. Definitely not. We have already had wells that have gone dry for portions of the summer in Patagonia and Sonoita. If the drought that we have been in for the past 5 or 6 years doesn't break, our water problems are just starting. Right here in Santa Cruz County we are doing better than a lot of the rest of the state. I just got off the phone with a friend who told me that Lake Powell is down about 140'.

What to do about the water problems is a little less black and white. If you are in the planning or building stages of a house, make sure you include a gray water system in your construction. Gray water is water that has already been used once for baths, showers, or laundry and can be used again for irrigation. Many washing machines use around 40 gallons per load. That can keep several trees alive for a while. If you are already built, see if you can knock a hole in a wall to run a pipe outside and take advantage of your gray water. Right now I have a small orchard of about a dozen fruit trees that are watered on gray water with no supplemental irrigation.

If gray water isn't an option, and you still want to plant but can't water, I hope you like cactus and succulents. These are the only kinds of plants that can survive being planted this time of year without water. Anything that appears to have leaves on it, even mesquites, needs additional moisture to help it get established.

Most importantly, remember where we live and try not to be stupid about it. Putting ornamental turf in front of your house because it "looks good" is dumb in an arid area. Planting a weeping willow here, unless it is on gray water, is a bad idea. It is never going to look as good as the ones you left in the Midwest or Northeast. My vote for the biggest waste of water in dry country (and I practice what I preach) is washing your truck.

Maybe with more voluntary water conservation the water restrictions could have been avoided in Sierra Vista. Maybe not. It seems that I have written about 1000 columns saying that living in the Southwest is all about water. It is always about water. Make that 1000 and 1.

Good Questions

We had some more good questions at the nursery this week that I thought were worth passing on.

How do trees grow?

Most of the time, pretty slowly. Actually the question was really about if plants grow from the base or the top. Trees and all plants grow from the tips of the branches. A branch that is about 2' off the ground will always be about 2' off the ground. If you were to carve initials, in a heart, on a tree trunk 4' up and go back and look for them 20 years later, the tree should be much taller but the initials would still be 4' high. The question is would the initials be the same?

The leaves on my Arizona Ash are curled up and gnarly looking. What is causing this and is there anything that you can spray them with?

Curling, gnarly leaves on Arizona Ash are one of the rites of spring around here. Pick off a leaf and open it up. You will probably see fuzzy small insects inside. These are Ash Tree Whiteflies. The curling is caused by them sucking the juice out of the leaves. Spraying doesn't work because they are protected by the leaves and you can't hit them directly. If the tree isn't badly affected, just pick the leaves off and get rid of them. Most of the time the tree will just grow out of the problem after it gets hot. The only way to treat a bad case of Ash Tree Whitefly is with a systemic insecticide.

My (you fill in the blank) didn't come back this year. How come?

I've had this question asked about lots of different plants this spring including verbena, butterfly bush and fruit trees. In fact, it was asked often enough that I decided to make up an answer that might even be correct. What made this winter different was the cold. Not how cold it got but how it got cold. It was fairly mild up to around Christmas. Two days later we got some of the coldest temps we have seen in a while.

Nothing had a chance to "harden off. Lots of the plants that were killed off have the ability to make though single degree temperatures but just couldn't take the rapid change. It kind of reminded me of the Great Freeze of '78 when the same thing happened but to a larger extent.

What is the "best" tomato for our area?
Because our summers are all so different there is no "best" for us. If you have a variety that has worked well for you over a period of years stick with it. The past few years Early Girl and Fourth of July have performed well for me. I will plant them this year for my production plants. For entertainment value I'm going with some heirlooms like Mortgage Lifter, Constuluto Genovese, and Burbank. It is always fun to see how they turnout.

Speaking about "turning out", it kind of reminds me of a conversation I overheard between two old farmers. One asked the other how his taters turned out this year. The second answered they didn't turn out at all, he still had to dig them out. If you have more questions and would like more quality answers like this, just stop by and we will see what we can do.

The Future, Part I

(Jim's notes: The conclusions that I am drawing are my "best guess" as to what is happening or going to happen should the drought continue. If someone wants to disagree with the predictions, or claim that I don't have enough hard science to back up my ideas, they may have a point. I still think I'm right, though.)

We are in about the 8th year of the drought. There are isolated pockets that have done OK in this time period but overall much of the Southwest is suffering. Most of the significant droughts of the last hundred years, including the droughts of the teens, twenties and fifties, lasted about 5-7 years. Megadroughts in the last 500 years have spanned 25-40 years. We are already seeing changes in the natural vegetation, and should the dry spell continue, things may get a whole lot different. Of course, as the plant life changes the animal life around it will have to adapt and change, too. Subtle change is always happening but here we could be talking about wholesale changes in a relatively short period of time.

The amount of change to the plant life will be in direct proportion to the elevation. Low desert plants (saguaros, Palo Verdes, etc.) are built to get by with very little moisture. Higher elevation species like pines and spruce can't. They don't store water well enough to survive long periods of minimal moisture. The lack of wet stuff leaves them stressed. In this weakened condition they can't fight off pests and disease. I have been told that the entire Blue Spruce population on Mt. Graham has already died off. Drought also promotes wildfires in the higher elevations. Most of the biggest wildfires in Arizona's recorded history have occurred in the last 5 years. Wildfires bring perhaps the swiftest changes.

In New Mexico it is estimated that 11,000,000 pinions have died off due to drought-related stress. I believe that number alone qualifies as a major change in vegetation. They will be replaced by something, probably native grasses and more drought tolerant shrubs.
There are changes happening on a local level, too. A few years ago a neighbor called me over to look at a population of Chihuahuan Pines that were dying on his property. This was a mature stand at about 5100' or so. Ordinarily this plant is happier at the higher elevations here. There was sign of beetle damage on the dead trees. The pines couldn't fight off the beetles after being stressed. I told the homeowner that he was seeing a population change in response to the climate. At the time I thought it was kind of an isolated event. Guess I was wrong about that.

There is a slope I have been watching closely in the base of the Huachuca Mts. for the past 6 years. This slope is southwest facing and rocky, which are both traits that would amplify the effects caused by a lack of water. This hillside is about three quarters of an acre in size. I did a survey on it the other day. This area contained about 129 mature trees. Most of them are Emory Oak with the rest being Alligator Juniper and 3 pinions. Out of the 129 trees I counted 28 that are dead or dying (won't make it another year). Almost all of the dead trees are oaks. In this one surveyed area we have lost 22% of the mature trees. Of course this is a selected area. Some are worse while other locations show little apparent damage at this time.

It may be more serious than I am figuring. In the surveyed area, only the trees that were in really desperate shape were counted as dying. These were trees that had already died back about half way and only had leaves growing from a few branches. I didn't count trees that that had a full set of leaves as dying even if they didn't look right. Maybe I should have. A neighbor called me over to look at a tree that had blown over in our last so-called storm. It was an Emory Oak with about a 14"inch caliper. At least 10" of the 14" had been completely consumed by a dry rot fungus at the base. The tree was hanging on by a couple of inches of decent wood. The amazing thing is that the tree looked as good as most of other oaks growing in that area.

Speaking about the way a tree "looks" is the single most discouraging sign out there to me. None of the oaks have any new growth on them this year. They put on a set of leaves and stopped. When I look back at the pictures of local Alligator Junipers from years past they are all thick and bushy. Each one made a good solid visual screen. Now they are more like filters. Houses that lay hidden in the past are now visible from the road.

Is it all gloom and doom? And what will it look like if we do head into a mega drought? We can look into that next time. One thing for certain, as always in our parts, it comes down to water.

The Future, Part II

(This was part of a series of articles talking about the visible effects of the drought on the native plant populations.)
The real trick right now isn't guessing if the drought will have a visible effect on our native plant landscape. That's already happening. Figuring out how much of a change is the hard part. Are the so called

"Sky Islands" (high elevation forests in the Southwest) on the way to becoming "Desert Islands"? Of course that depends on if the drought is broken, or if what we are calling a drought becomes the norm for us in southeast Arizona.

Being as we are making this up as we go along, let's say that for the next 5-10 years we continue to be below historical average for moisture by about a third. You would still recognize the countryside but it might look a lot different. The typical oak woodlands would look more like grasslands that a fire had passed through. Lots of standing dead trees. Mesquites might start to move in and establish stands as they get by on less moisture than the oaks. The north facing slopes and canyon bottoms would still hold some relic old trees.

What we see now as grasslands would take a turn towards the "deserty" side. You could expect more plants like acacia, prickly pear, cholla and burrow weed. No palo verdes or saguaros, though. In our scenario it has gotten dryer, especially in the winter, but not warmer. In fact you could expect some real cold nights because there would be less clouds to blanket in the heat. The hard freezes are a limiting factor on where a lot of the low desert plants can grow.

We could go on describing all the changes to the different plant communities, but you can do that yourself. Just imagine most everything resembling the plant groups from the same locations and exposures that are 1000' to 2000' feet lower in elevation.

Now at this point you are probably saying "Ol' Jim has really lost it this time. There have always been dry spells, wet ones too. Changes like this don't happen." Well, folks, they already have, just not in our lifetimes. According to my pal Paul Martin, towards the end of the last Ice Age the areas that are now oak woodlands were pine forests and the current ponderosa pine forests were dominated by spruce. There is a gradual desertification that has been taking place for the last 10,000 years. The hard question is: do major changes take place over a couple of decades, or does it take hundreds or thousands of years. The answer to both questions is probably yes.

Hopefully our drought starts to break tomorrow and I am wrong about all my guesses. Then it would be back to the good old days of trucks getting stuck in the mud, school being closed because of snow, hillsides washing out, and lots of weeds. Oh, for the good old days, like 10 years ago.

Good Bye Old Pal

We sure lost a good one a couple of weeks ago when Wayne Wright passed on. For those of you that didn't know him, or only thought of him as the "old timer" that always waved, you missed one heck of a man, cowboy, and story teller. Wayne was a real one. Horses were tools to get a job done, not pets. Progress was good because it meant prosperity. The weather was gonna do what it was gonna do.

No matter who you were or what you did, if you worked hard, Wayne respected you. And hard work was what Wayne's life was all about. From catching wild horses as a young man to cutting, delivering, and stacking hundreds of cords of wood ($20 a cord), to shoeing horses ($1 a head), he was a working man. One time Wayne had a special lift built so he could continue to shoe even though his leg was broken.

And once when he was a kid, an aunt lost about a thousand sheep in a blizzard in Wyoming. When it thawed that spring, he and his cousins had to go rake the wool off the dead sheep to salvage what money they could.

Wayne thought it pretty silly if you tried too hard to be a cowboy. One day, a person that had just moved to town was leaning against the wall of the feed store. He had his cowboy jeans tucked into his shiny new boots. Wayne walked by and said to me (not caring if he was heard) "That feller seems to think he belongs here."

Above all, in his later years, Wayne was a story teller. He cowboyed all of his life and had a sharp eye for detail. When he shared stories about growing up in Wyoming, working on ranches in Montana, or running sheep and cows in Santa Cruz County for the last 50 years, you were given a special view into a life most of us never lived. How about riding in a 100 mile race as a young boy against a specialty horse that traveled from town to town and took on local horses? Wayne won it on a big gray (part draft horse) in about 10 hours. The horse just took off on a trot and never quit. Wayne spent the night in the stables and rode 100 miles home to work the next day.

Hard to say what my favorite story that Wayne told was. It might have been the one about working in a cow camp in Montana in - I am guessing- about the late 40's. Wayne told me there was a kid that insisted on carrying a six shooter (all of the experienced hands carried rifles). A pistol shot was heard in the distance, so the foreman rode over to check it out. When he got to the kid he asked "What happened?" The kid told him "I shot a rattle snake." The foreman answered "What the hell did you do that for? It might have bit a sheep herder!"

Of course, as with any good story teller, Wayne had a bit of practical philosopher in him. "In winter, fat is the best color for a horse."

When Wayne was in his 70's, he and Rory George were working some of his cows. A rank steer bolted through a steel gate, flinging it open and crashing it into Wayne's head. He lay on the ground motionless and bleeding from places where blood shouldn't be coming from. Rory thought that it was the last of Wayne, who was airlifted to a Tucson hospital. Now, Wayne was plenty familiar with hospitals, having broken most every bone in his body in some kind of horse wreck. About 3 or 4 days later, he was home saying that he needed to get out of there because the doctor wanted to do plastic surgery on him and that he wasn't about to make that doctor's boat payment. Clem made him give up the cow business shortly thereafter.

Clem? Clem was Wayne's wife for over 60 years, and they were a team. One time, Wayne came in the nursery to get some flowers that Clem wanted. I told him that I would give him the flowers, but I know that Clem wouldn't like that so I suggested that he give me a dollar. That way, when Clem asks if he paid for them he could honestly say "yes." A few days later Clem came in - shaking her finger at me - and saying that I didn't charge enough for the flowers. I waited until she left and said to Wayne "You have been married for over 60 years and you can't get away with $3 worth of flowers?" He just smiled and said, "guess not" which I took to mean he wouldn't have it any other way.

The Memorial Service for Wayne took place at the Santa Rita Abby. Most of us had gotten rain the day before and there were good looking clouds in the sky. That was perfect because weather was always one of Wayne's favorite things to talk about. The service was moving. The Sisters really do sing like angels. I just wish there was a part where we could have all stood up and shared our favorite story about Wayne. On the other hand if we had done that, we might still be there.

Mental Heat Rash

(This was actually put together at the start of August 2005. I found it in an old note book and after reading through it, it is obvious that I must have been suffering from a Mental Heat Rash when I wrote it.)

We finally did it. July '05 was officially the hottest month ever on record. The heat has settled into my brain and has got me wondering and thinking about some of the things I have learned working with plants and people the last 30 years. Right now…

I wonder if we are in a drought or if this is just the way it is going to be for awhile.

I wonder why anyone would want green grass here in the Southwest year round. Don't waste the water. You appreciate it so much more when it happens for 8 weeks a summer.

I wonder what the heck was so wrong with VCR tapes (that I could play) that we had to replace them with DVDs (that I have to wait for someone to start for me). It seems to me the movies on them haven't changed all that much.

I wonder why anybody prunes anything into a geometric shape.

Pretty short "wondering" list isn't it? Good thing life isn't all wondering. Here are a few things I have learned.

I have learned a rose is prettiest before it opens all the way. (There might be a moral there but it is way too deep for me.)

I have learned that the more I make myself slow down while working, the faster the job gets done.

I have learned that you will never, ever reach an irrational person with a rational line of thought. People are going to believe what they want to believe.

I have learned that when someone asks you "so what do you think about" (a fire district, for example)? What they really mean is that they want the chance to tell you what they think about it for the next 15 or 20 minutes.

I have learned that it's never a poor man that says "money isn't important".

I have leaned that just because a joke contains the words "redneck" or "duct tape" that don't mean that it is funny.

I have leaned that dry heat still cooks a turkey.

I have learned that talking to plants doesn't do squat for them. Watering does. If you don't believe me take two plants. Water one. Talk to the other.

I have learned that life is too short to buy cheap tools or bad musical instruments.

I have learned to plant landscapes sparse but pack color into pots and planters.

I have learned that the larger a person's "forward" list is on their e-mail, the less friends they probably have.

I have learned that the problem with blowing smoke is that, when it clears, people can see through it. (Actually, I haven't really learned that one yet but I am trying.)

And finally, while walking through Wal-mart one particularly hot and sweaty morning

I have learned that if your belly is sticking out farther than your breasts, you probably shouldn't be showing off either in public.

I have heard this is true for the women shoppers, too.

Epilogue:

Adios

About ten years ago I decided to open up a retail nursery in Sonoita, AZ. Naturally I went into the nursery business thinking I knew way more than I actually did. After all, I already had almost twenty years of working with plants in the Southwest. Soon, however, it became obvious I still had much to learn about the growing conditions, plants, and especially, dealing with people. The learning was fun and after a few years I had a handle on the nursery business. You always remember the first sale and my first one was to a lady, in a station wagon, that was just making a u-turn in the parking lot. I ran out and asked if she needed anything. Looking back, I am guessing that she was glad I wasn't trying to wash her windshield for spare change and she bought a 5 gallon red yucca.

I believe I was one of the few nurseries left in the country that actually sells plants, and only plants, and doesn't try and pass itself off as a "Garden Center." Recently the nursery had begun to do pretty well. It made a living for my family, the work wasn't too hard, and I had time off in the winter. After a decade I had accumulated a fair amount of working knowledge about what and how things work in the higher elevations of the Southwest. Obviously, with things going this smoothly, it was time to get out and try something else.

Anyone who has been involved in a retail business will tell you it is the people that make or break it. We were really lucky because almost every day we were visited by good folks who appreciated the service and knowledge and would put up with the lack of organization. When you deal with people there are always going to be memorable encounters and two of my favorites from the last ten years were the time someone told me not to plant tomatoes next to chiles or you will end up with hot tomatoes (I wish this was true as it would prevent a lot of chopping for making salsa), and the time someone asked if they could "borrow a Christmas Tree, just for Christmas Eve", but promised they would bring it back on Christmas Day. Unfortunately I was already sold out for the year but I probably would have let them do it. The all time favorite probably goes back to when I was landscaping in Tucson some twenty years ago. One person told me that the soil in the planter area around the front door always has to be removed in new construction because "that is where all the construction workers urinate."

One of the biggest benefits of being in a retail business is trading. During the last ten years trades were made for plants, fossils, mineral specimens, goat cheese, signage, rhubarb, welding equipment, wine, a truck, various produce, dinners at a local eatery, a giant pumpkin, peaches, beef, scrap metal, pet food, music lessons, graphic illustrations, and thanks to my friends the Hersey family, chickens. When the chicken deal went down I knew it didn't get any better than this and it was time to move on to a new line of work. Swapping will be one of the things I miss.

A fair question now would be "why get out when it is all good?" Good question. And for a change I will give a serious answer. Family. Now "family" can be used a couple of different ways. When a politician has just gotten caught in a broom closet with two exotic dancers, a blind Great Dane, and a chicken, he might say he is resigning to "spend more time with his family." I am happy to report that isn't what I mean by "family." I have spent the last two Saturdays fishing and playing basketball with my kids. That is my definition of family. Saying that life is too short is no great insight. Trying to do something about it might be. The fact that I turned 50 last fall, I am sure, had absolutely nothing to do with this decision.

So now it is on to reseeding, revegetation, and erosion control projects. It is good work and I enjoy being involved with it. Of course I will have to make time for the demands of being a famous author, too. I am going to try and not work too many weekends but if we get real busy I will just make the kids work. That will be some good family time. Thanks to all who made the last ten years at Diamond JK Nursery enjoyable. Oh yeah, I almost forgot, if anyone has anything good to swap for native grass seed or reseeding, I can be found at Arizona Revegetation and Monitoring Co.

Special Bonus Section: Reseeding; The Art of Bring it Back

One of the best projects you can take on is the restoration of damaged land. Improvements will slow down erosion, improve wildlife conditions, allow better water penetration into the ground, increase value, and most importantly look a whole lot better. Unfortunately much of the discussion of land stewardship has been high jacked by extreme environmental types. Conservation is a good thing and is for each and every one of us to practice.

SYS – SAVE YOUR SOIL   p. 167

SEEDING BASICS   p. 168

WHAT'S LOVE GOT TO DO WITH IT?   p. 170

RANGE RESTORATION – GETTING IT DOWN   p. 172

PASTURE RESTORATION   p. 173

WILDFLOWER PLANTING   p. 175

SYS – Save Your Soil

Recently, I traveled to the exotic locale of Long Beach, CA to attend the International Erosion Control Association Conference. The training I was there for had to do with establishing native plants for soil erosion control. As a seeding contactor I wanted to learn better methods of ground preparation, seed mix calculations, and ways of application. The training touched on all of the above, but what I really took from the class was that the most important part of land reclamation happens way before the seeding process. Save your topsoil.

Now before those of you of Midwestern origin laugh and say "show me some topsoil", let's define what topsoil is here in the Southwest. For our purposes topsoil is the small layer of dirt above the caliche, clay, or rock, where the native vegetation grows. In most cases it is about 2" – 12" deep.

The main reason why it is important to the process of bringing back disturbed ground is that it contains a "seed bank" of native material that is suited to growing in your exact location. Also our topsoil has a small amount of organic material in it which is lacking in the subsoil. Finally, our soil on top contains microorganisms (bacteria and fungi) that help the specific plants in our area get established and grow.

Microorganisms are one of the hot topics in the plant-growing world. Seems like everybody is trying to hustle them as a miracle cure for what ails your plants. One of these days I expect to see unwanted ads start to pop up on computers "buy real herbal micronutrients without a note from your local Agricultural Extension Agent, free trial offer." Being a skeptical son of a gun, I don't believe that cheap imported micronutrients from China are going to do us one bit of good. However, research has shown that local, to the specific area, micronutrients are very important to plant health. Microorganisms could and probably should be a whole column by themselves, but basically I don't know a lot about them yet so that is going to have to happen at another time.

The best part of saving your soil is that it is easy to do. Just put aside this layer as part of the grading process. As most dirt movers (not exactly known for being the most progressive of individuals) would be new to this idea, it may be something that you have to insist on and could add to the cost of the dirt work. The little bit of additional expense would be more than made up for in a better revegetation job. Before this soil is replaced, the existing dirt should be ripped to allow water penetration and allow the two layers to mix. Saving your soil is going to be the quickest and best way to help restore your land to the way it was before it was disturbed.

Seeding Basics

Here are a few questions that we have been getting asked lately about the reseeding process.

Q. When is the best time to put down native grass seed?
A. Are you sure that we can't start with an easier question? I am not sure there is a right answer to this one. In the natural world seed drops in the fall. The theoretically gentle winter rains help to settle it into the soil. Fall is a good time to seed. On the other hand seeding just before the summer rains means the seed will be on the ground for the shortest amount of time before it can germinate. This would lessen the amount lost to wind, insects, and birds. So, early summer is also a good time to seed. As a seeding contractor we seed year

round and slightly adjust the seed mixture to make the most of whatever season it happens to be. I guess the best time to seed is right before an extended wet period.

Q. What conditions are necessary for good germination of our warm season native grasses?
A. Good seed + 80 degree or higher ground temperature + 8-14 days of constant moisture on the seed bed = germination of our warm season native grasses.

Q. What do we have to do to prepare the ground before seeding?
A. Unless the ground has been slicked off and compacted don't do too much. The looser and rougher the surface the better. You don't have to remove the rocks as they hold moisture and give the seed someplace to hide. Every depression or pocket holds moisture and give the seeds a better chance. Compacted dirt just sheds water and dries out too fast. Disturb it so the water slows down and can penetrate into the ground. Do whatever you can to rough it up with tractor rippers, a pick, or even with a rake.

Q. What is the single most critical part of the reseeding process?
A. Moisture. You can always add more seed and you know that the ground will eventually be warm enough, but moisture is the wild card. Irrigation is a great equalizer but it isn't always practical in an arid area. The trick is to get the precipitation that does fall to stick around for a while. That is why soil preparation is important. There are also a few low tech methods that can be used for slowing down water run off. Start by laying branches or rocks in a line perpendicular to the flow of the water. A light mulch will also retain some moisture and help tip the odds in your favor. A steep slope, especially south or west facing, will be more difficult than flat or gently sloping sites to reseed.

Q. How long does it take an area to come back after it is reseeded?
A. Relax, my impatient friend. It took many thousands if not millions years for our part the world to get the look that it has. Then you

scraped it clean in about an hour and a half, or left 4 horses on 1 acre for 5 years. It is not coming back in 6 weeks. Most jobs take several years before they look "natural".

Q. What is the best species of grass to use in Southeast Arizona?
A. Never limit yourself to a single species. If you choose wrong you just wasted a bunch of money and effort. The shotgun approach works best. I like a 6-8 species mix. Some grasses will prefer different locations even on the same piece of property. A mix gives you the best chance of having a species for every situation. Your mix should also include species that germinate quickly and act as pioneer plants. These help slow down water, break up the surface, and give shelter to the longer lived but harder to germinate species.

Q. Speaking about species, what about Love?
A. I am sure glad you asked. If you have the time I would like to share a few details about how lucky I have been in this wonderful experience. You see it all started when…

Q. No, you nimrod, I was asking about Lehman's Lovegrass, not your personal life.
A. Sorry that is a long story. I guess that will have to wait until next time.

What's Love Got to Do with It?

(We left off the last column about the basics of reseeding with a question concerning lovegrass.)

Q. What's Lovegrass got to do with reseeding?
A. Take a drive south of Sierra Vista, or west of the Mustang Corners on Highway 82, or through Elgin. Look at all that beautiful native grass growing on both sides of the road. Looks pretty good doesn't it? First of all, looks are sometimes deceiving and secondly most of what you see isn't native at all. It is an African import, Lehman's Lovegrass.

Back around 1932 some folks in the US Soil Conservation Service introduced a species of grass that they figured might do well here. This invasive exotic species has taken over large parts of SE Arizona. But, as usual, there are two sides to the story.

The biggest problem with Lehman's Lovegrass is that it replaces the native species and forms huge stands of nothing but Lehman's. This is bad for a few reasons. Animals, birds, and insects depend on a variety of native grasses for food. This monoculture limits the food sources. Lehman's also burns at a hotter temperature than native grasses. The trees and shrubs that have evolved to withstand "native" fires may be destroyed by the higher temperatures of burning Lehman's. Finally, the problem with any pure stand is that should something come along, like a disease which targets that species, you will be left with nothing. Most live stock will avoid Lehman's, except when first growing, if there is any other choice.

Q. Boy, this stuff is really evil and bad for the rangelands, right?
A. If only life could be that black and white. Back around the turn of the last century a lot of this area was in poor condition because of a couple of droughts and mismanagement of livestock. Lehman's germinates easier, and responds to fire better, than most native species of grasses. These qualities helped it get established quickly and hold soil in place that might have blown or washed away. Land that is covered with Lehman's sure looks better than the ground I have seen with no grass cover by the border. The mesquites that grow there look like they are up on pedestals of a foot or more because any soil that wasn't held by their roots has eroded.

Q. Getting back to reseeding (finally), is it a good idea to use Lehman's as part of a seed mix?
A. No. For the same amount of time, effort, and money you have a chance to practice some good conservation and restore some of our grassland. This is a good move to help local wildlife. Besides, if there is any Lehman's in the area it will come back on its own. If there isn't, you don't want to be the one that introduces it.

Q. Why don't we just get rid of it?
A. Lots of research has been done on that. Like its other African cousins Bermuda and Johnson Grass, this species is here to stay. Estimates are that Lehman's seed may still be able to germinate after 40 years on the ground. Most of our native grasses have a viability of about 10 years or so. It is not going away. We need to learn how to manage it.

Range Restoration - Getting it Down

There are many different ways for getting seed on the ground. The good news is that providing you meet the formula (good seed + proper ground temperature + adequate moisture) they can all work. The trick is to figure out what method is the best and most economical for the job you are doing. Like most jobs a little effort beforehand can save lots of work later on. The most important thing to do ahead of seeding is ground preparation.

Anything that can either slow water from running off, or offers protection for the seed, is good. On construction sites or horse pastures the ground is often slick and compacted. Break it up. Rocks are good. Leave them where they are. Any ripping or raking should be done perpendicular to the direction of runoff. The object is to create "safe sites" or little pockets in the ground that can hold moisture and give the seedlings a better chance of taking hold.

The cheapest way to get seed on the ground is to hand broadcast it. Much of our native grass seed is too light and fluffy to go through hand seeders. It just hangs up. Blue Grama, Sideoat Grama, Cane Beardgrass, and many more will just not go through these devices. Very technically speaking this hand broadcasting method is referred to, by range management specialists, as "feeding the chickens". Hand seeding is usually done on smaller areas. It also can be used on conditions where you are trying to supplement or add species to an existing patch of vegetation. A good general seeding rate for most types of seeding on bare ground is ½ lb to 1 lb per 1000 sq. ft. or 20 lbs. to 40 lbs. per acre. This makes the assumption that a general mix of native grasses is going down. Of course, each individual species has its own rate of application. The biggest drawback to hand seeding is that it doesn't give very good seed to soil contact which is necessary for good germination. You can help settle the seed by watering or lightly tamping it after broadcasting. Hand seeding probably has the highest loss of seed due to wind and seed feeders like birds and insects.

Hydroseeding is the application of seed, mulch, and something that makes it stick. It is shot out of a tank under pressure. Almost any kind of seed can be applied this way. This is "the green stuff on the side of the highway" that you see along highway construction sites. The mulch helps hold moisture for the seed. It also locks up the seed so the loss to birds and insects is reduced. Hydroseeding works well on construction scarring and leach fields. It is also used extensively on slopes. This method of application can also provide some dust control. Hyrdoseeding is not something the homeowner does by themselves as it requires specialized equipment. It works well on medium sized jobs, usually less than an acre. There are more cost efficient ways for seeding larger portions of ground like your 5 acre horse pasture.

Land Imprinting is fairly new on the reseeding scene. It was developed by Dr. Bob Dixon of Tucson around 1976 for arid lands reclamation. Basically, imprinting is a process that creates pockets in the ground for the seed. These pockets hold moisture and reduce the wind on young seedlings. Land Imprinting is often a good method to bring back your hammered pasture. Now hold your horses, we will get into imprinting and managing your pastures next time.

Pasture Restoration

Short of scraping off the top 6 inches of dirt, one of the worst things you can do to your land is put more horses on it than it can handle. And it can probably handle less than you think. I asked two of the most knowledgeable fellows in all of Southeastern Arizona, Dean Fish and Mac Donaldson, what they thought the ratio of acres to a horse should be. Dean said that in the best grasslands in a good year he

thought about one horse per 40 acres. Mac thought that under average conditions it takes 80 -120 acres per head!
Of course, supplemental feeding helps but remember horses are like people. They don't have to be hungry to eat. They will graze out of boredom and the result is the same. With supplemental food they just poop more. It is all about management. So how is the best way to manage 5 acres with a few horses?

In my humble (sometimes) opinion the most important thing you can do is keep them confined to a very small portion of it. Let them out to graze occasionally but monitor how hard the pasture is being hit. This means that you will be more responsible for working them or getting them exercise. If you can't exercise them and want your land to be in good shape, then sell them, give them away, or eat them. Get a goldfish for a pet. If the land isn't already hammered, here's something else to try.

 About 10 years ago I had a pair of horses on a small pasture of less than one acre. As soon as the summer rains started I would remove them and keep them off until the grass had gone to seed. This rested the pasture in its critical growing period from July to September. Then we would put them back on. Every year we went from basically bare ground in June to mostly grass coverage by October. The species changed from predominantly Blue Grama and Sideoat Grama to coarser grasses such as Green Sprangletop, but more importantly we weren't losing dirt to erosion. The property was kept in decent shape.

If your ground is already looking like the parking lot at the county fair grounds it is going to require two things. You have to open it up from all the compaction so the moisture will be able to penetrate. Also, seed needs to be added so a new crop of grass can get started. If you just rip or disc it and don't add seed, you will probably end up with a field of weeds.

Land Imprinting or seed drilling are two of the best and most cost effective methods of restoring pastures. They accomplish both the goals of opening up the ground and reseeding. Imprinting, if properly done, creates offset pockets in the ground which are about 1 ft. sq.

and 4"-7" at the deepest point. These pockets keep the young seedlings out of the drying winds, gather organic material that would have just blown by, and most importantly, collect rainwater. In fact, because of their shape they actually multiply the effect of the water that does fall.

Whatever method of restoration you go with, remember that it is going to take time. It took thousands of years to get the land to look like it does here in Southeast Arizona. We are able to mess it up (mismanage) in a couple of hours with equipment or a couple of years with livestock. Dr. Bob Dixon, the man that created the imprinting process, suggests a five year period for the revegetation to really take hold.

Finally, here is a question I often get asked by people that have just moved here from another part of the country where grass is actually green for more that 8 weeks a year. "We just bought 5 acres and got a well so water isn't a problem. What can we plant here that will feed our 4 horses?" Well, there is only one kind of "grass" that will produce enough to feed 4 horses on 5 acres. And it is illegal. Go build a corral and buy some hay.

Wildflower Planting

Unlike the low stinking desert where, if they are lucky they get to see a few Poppies in March, we have two wildflower seasons. They follow the winter and summer moisture periods. If you are trying to encourage wildflowers on your property, try to seed at the start of either season. Just like the native grasses using a mix will give you the best chance of having some success. Wildflower planting can be kind of tricky. Good thing we have two distinct paths that we can choose to take. Do it, or do it right.

Do it: Take a handful of seed and just scatter them wherever you want wildflowers. Then hope for a really wet year. Personally, I keep trying this method and have come to the conclusion that it just doesn't work very well. It is totally dependent on the timing of winter moisture, which is good about once every 8 — 12 years. Last year was real good, so the odds are that it will be a while before it happens again. If you are patient and not too ambitious, this might be the method for you.

Do it right: Although almost all instruction for planting wildflowers varies slightly, they can be boiled down to this. Loosen the top few inches of your soil. Remove all weeds or unwanted plants. Spread the seed and lightly cover with soil, no more than one half inch. If you are working with native or southwest-adapted seed no soil amendments are needed. Water enough to keep seed moist. Maintenance is limited to watering and weeding.

In Southeast Arizona we are able to grow both cool and warm season wildflowers. Your best choice of seeds for an area is always a mix instead of a single species. The more types of flowers the better. This increases the chance that some will like your specific area and naturalize. A cool season mix could include the poppies, flax and Arizona Bluebells. The Mexican Poppy isn't quite as showy as the California type but lasts much longer into the spring. For the warm season, gaillardia (Indian Blanket), coreopsis, Mexican Hat, and penstemons all grow well. I gained a lot of respect for the globe mallow two winters ago in the drought. They were the only wildflowers that bloomed that spring.

A good wildflower mix should cost about seven to ten dollars an ounce. That amount covers about 200 sq. ft. Watch out for "native" mixes that are native to other regions or different parts of the country. They won't work as well here.

With a good start the first year some of the species should come back year after year on their own. Of course that depends on the amount of moisture we're getting that particular season. That's the way it is for just about everything in the Great Southwest.

Subject Index

Africanized Bees, 53
Agastache, 103
Apples, 47, 48, 49, 51, 52
Arizona Cypress Borers, 59, 60
Bt, 16, 23, 26, 134
Caliche, 113, 114
Chiles, 146, 149, 150, 164
Christmas Trees, 82, 83
Drought, 4, 5, 6, 7, 8, 12, 17, 18, 29, 37, 41, 42, 47, 59, 60, 86, 87, 88, 90, 93, 101, 108, 110, 111, 116, 117, 118, 119, 121, 122, 125, 127, 137, 152, 155, 156, 157, 158, 158, 162, 171, 176
Fruit Trees, 1, 48, 49, 50, 51, 57, 58, 81
Glyphosate, 19
Gophers, 58, 135, 136
Grasshoppers, 44, 52, 54, 56, 57, 58, 61, 93, 134, 137, 147
Groundwater, 6, 7, 142
Harvester Ants, 20, 21
Heirloom Tomatoes, 147, 148, 149
Hummingbirds, 14, 15, 22, 54, 66, 67, 94, 96, 98, 99, 103, 106, 119, 121, 127, 128, 145
Hydroseeding, 173
Illegal Immigration, 37, 66, 71
Iron Chlorosis, 77
Land Imprinting, 173, 174
Leafcutter Bees, 37, 143
Manzanita, 12, 13, 14, 29, 71, 108
Microclimates, 43, 44, 140
Mt. Lemmon, 27, 28, 29, 30
Native Plants, 14, 37, 74, 84, 86, 105, 115, 125, 167
Native Grasses, 1, 10, 84, 94, 102, 136, 156, 169, 171, 172, 175
Palmer's Drought Index, 41, 42
Peaches, 37, 49, 50, 59
Penstemons, 1, 14, 15, 66, 85, 121, 176
Pigweed, 18, 106

Planting, 8, 12, 17, 18, 31, 44, 45, 48, 61, 76, 77, 83, 84, 104, 105, 108
Planting Dates, 140
Pruning, 51, 55, 58, 61, 80, 81, 82
Rabbitbush, 45, 46, 119, 120
Revegetation, 165, 168, 175
Rhubarb, 46, 47, 48, 164
Round-up, 19
Salvias, 46, 84, 99, 102, 116, 118, 120
Sapsuckers, 58
Soil Preparation, 31, 123, 169
Soils, 77, 88, 106, 114, 135, 136, 142
Sudden Oak Death, 90, 91, 92, 93
Summer Rains, 1, 7, 8, 20, 56, 95, 101, 123, 141, 142, 168, 174
Tobacco Hornworm, 22, 24
Tomato Hornworm, 16, 22, 134, 147
Tomatoes, 16, 23, 37, 88, 89, 95, 140, 147, 148, 149, 164
Tree Tomato, 89
Water Plants, 131, 132
Watering ,17, 18, 67, 89, 90, 93, 95, 111, 114, 118, 124, 151, 152, 163, 173, 176
Wattles, 35, 36
Wildflowers, 85, 109, 120, 123, 141, 175, 176
Wind, 5, 6, 9, 11, 36, 43, 44, 48, 79, 80, 86, 93, 100, 103, 111, 112, 113, 116, 118, 125, 125, 141, 143, 148, 168, 173, 174

For more information on...

Dana Cude, Artist twodogs@theriver.com

Jim Koweek, Author www.azreveg.com

Leonard Sadorf, Sonoran Wind Press sonoranwind.com

Zackery Zdinak, Artist www.lifedraw.com

Peter Gierlach, "Petey Mesquitey" can be heard on "Growing Native" on 91.3 FM KXCI in Tucson, AZ. www.kxci.org

List of Illustrations

By Dana Cude:

Summer Rain, P. 4
Devil Grasshopper, P.56
Hot Rocks, P. 104

By Zackery Zdinak:

Sacaton, p. ii
Longhorns, p. iv
New Mexico Spadefoot, p. ix
Sideoat Grama, P. 2
Velvet Ants, P. 9
Manzanita, P. 13
Harvester Ants, P. 21
Mohave Rattlesnake, P. 33
Sacaton, P. 38
Rubber Rabbitbrush, P. 45
Blue Grama, P. 72
Apache Pine, P. 87
Purple Cone Flower, P. 96
Mexican Elderberry, P. 100
California Poppy, P. 122
Tarantula, P. 138
Desert Willow, P. 145
Pine Snag, P. 158
Tobosa, P. 165
Cryptobiotic Soil, P. 167
Giant Hairy Scorpion, P. 179
Diamondback Rattlesnake, P. 180

www.ingramcontent.com/pod-product-compliance
Ingram Content Group UK Ltd.
Pitfield, Milton Keynes, MK11 3LW, UK
UKHW051255180426
11947UKWH00020B/1731